SpringerBriefs in Environment, Security, Development and Peace

Volume 35

Series Editor

More information about this series at http://www.springer.com/series/10357
http://www.afes-press-books.de/html/SpringerBriefs_ESDP.htm
http://www.afes-press-books.de/html/SpringerBriefs_ESDP_35.htm

Larry D. Barnett

Demography and the Anthropocene

Springer

Larry D. Barnett
Professor Emeritus
Delaware Law School
Widener University
Wilmington, DE, USA

More on this book is at: http://afes-press-books.de/html/ESDP_35.htm

ISSN 2193-3162 ISSN 2193-3170 (electronic)
SpringerBriefs in Environment, Security, Development and Peace
ISBN 978-3-030-69427-2 ISBN 978-3-030-69428-9 (eBook)
https://doi.org/10.1007/978-3-030-69428-9

Copy-editing: PD Dr. Hans Günter Brauch, AFES-PRESS e.V., Mosbach, Germany
English Language Editor: The author

This Springer imprint is published by the registered company Springer Nature Switzerland AG
The registered company address is: Gewerbestrasse 11, 6330 Cham, Switzerland

To

Linda Rae

Acknowledgments

This work was made possible by the online libraries at Widener University and the Delaware Law School. I want to express my appreciation to the law school library staff for obtaining materials that were not in the online libraries.

The series editor, Dr. Hans Günter Brauch, and the referees who evaluated the manuscript warrant mention, too. Each invested considerable time in my work and was remarkably helpful to me. I a m grateful for their efforts.

Wilmington, DE, USA
December 2020

Larry D. Barnett

Contents

List of Figures

List of Tables

Chapter 1
The Population Factor

"[W]e have always assumed that progress and 'the good life' are connected with population growth. In fact, population growth has frequently been regarded as a measure of our progress. If that were ever the case, it is not now."

The above passage was penned in the early 1970s,[1] but human overpopulation is an idea with a very long history—indeed, the idea appears in works written during antiquity.[2] For example, Tertullian, an influential theologian from the Roman city of Carthage who lived during the second and third centuries A.D., wrote that "the most important evidence that the population is so great is that we are now a burden on the World, there are barely enough of the essentials for us, … and … nature can no longer sustain us."[3] An even-earlier narrative describing an instance of overpopulation is provided by the Greek historian Herodotus, who lived in the fifth century B.C. In his work *Histories*, Herodotus wrote that, due to famine-caused food shortages among the Lydians, "their king divided the whole Lydian people into two parts, and he appointed by lot one part to remain and the other to go forth from the land; …. So the one party of them, having obtained the lot to go forth from the land, … built

[1] The passage is from COMM'N ON POPULATION GROWTH & THE AMERICAN FUTURE, POPULATION AND THE AMERICAN FUTURE Ch. 1 (1972), https://issuu.com/js-ror/docs/720327_rock-pop. The Commission was created by an Act of the U.S. Congress; the Act was signed by then-President Richard Nixon on March 16, 1970. Pub. L. No. 91-213, 84 Stat. 67 (1970); *Remarks on Signing Bill Establishing the Commission on Population Growth and the American Future*, PUBLIC PAPERS OF THE PRESIDENTS OF THE UNITED STATES: RICHARD NIXON 266 (1971), https://quod.lib.umich.edu/p/ppotpus/4731750.1970.001?view=toc. The background, formation, and work of the Commission are described in Charles F. Westoff, *The Commission on Population Growth and the American Future: Its Origins, Operations, and Aftermath*, 39 POPULATION INDEX 491 (1973).

[2] Richard Harrow Feen, *Keeping the Balance: Ancient Greek Philosophical Concerns with Population and Environment*, 17 POPULATION & ENV'T 447 (1996) (pointing out the idea of overpopulation that is in the works of Plato and Aristotle).

[3] Arthur McCormack, *The Population Explosion – Myth or Reality?*, 45 LINACRE Q. 264, 264 (1978).

L. D. Barnett, *Demography and the Anthropocene*,
SpringerBriefs in Environment, Security, Development and Peace 35,
https://doi.org/10.1007/978-3-030-69428-9_1

ships for themselves, wherein they placed all the movable goods which they had and sailed away to seek for means of living and a land to dwell in."[4]

Today, an appreciable body of credible evidence (reviewed in *infra* Part 1.4) supports the thesis that the number of human beings who are now on the planet is close to, or may have already exceeded, the ability of Earth to provide adequately for *Homo sapiens*. Expressed more forcefully, the evidence indicates that growth in the number of people poses very serious risks to the human species and that this growth inflates the impact on the biological-physical world of other factors, e.g., income gains and consumption patterns. Yet oddly, neither the evidence that the current size of the human population has had material, negative effects on nature nor the continuing increase in human numbers impresses environmentalists or scholars in environmental law.

Perhaps, however, we should not be taken aback by this situation. After all, a major perceptual shift typically demands an "agonising reappraisal of accepted values, criteria of relevance, [and] frames of perception."[5] As a consequence, this chapter delves further into the situation that now prevails in environmentalism (Part 1.1) and presents a statistical picture of the magnitude of human-population growth over the last one hundred years (Parts 1.2 and 1.3). The picture, which attempts to convey the significant scale of this growth, pays particular attention to the United States because, compared to the inhabitants of other nations, U.S. inhabitants have a per capita impact on nature that is among the highest in the world. Chapter 1 ends with a detailed, evidence-grounded review of the multiple ways in which the natural world has been damaged by the expanding human population (Part 1.4).

Chapter 1, in short, aims to persuade readers that the increased and increasing number of human beings on the planet must not be ignored or taken lightly. Chapter 2 has a different focus. Specifically, it explores the extent to which Americans who worry about environmental quality express unease with ongoing numerical additions to the population of *Homo sapiens*. If the former is unrelated or minimally related to the latter, the attitudes of Americans generally may account for the avoidance by U.S. environmentalists of the subject of human-population growth. Prevailing attitudes in other countries may, of course, affect environmentalists in those countries. Lastly, Chapter 3 discusses the sociological role of law and government policy—their social

---

[4]HERODOTUS, 1 THE HISTORY OF HERODOTUS Book I ch. 94 (G. C. Macaulay trans., 2008) (1890), https://wps.pearsoncustom.com/wps/media/objects/2426/2484749/chap_assets/bookshelf/herodotus.pdf. Although the mass famine and mass migration recounted in Herodotus' narrative may not have happened, the narrative was obtained by Herodotus from a source, and since the subject of the narrative was a population that exceeded the available food supply, the source as well as Herodotus understood the idea of overpopulation and recognized the potential consequences of overpopulation. *See* Robert Drews, *Herodotus 1.94, the Drought ca. 1200 B.C., and the Origin of the Etruscans*, 41 HISTORIA: ZEITSCHRIFT FÜR ALTE GESCHICHTE [J. ANCIENT HIST.] 14, 15–17, 22, 29–30, 34, 37–38 (1992) (suggesting that the source sought to replace the explanation that was popular in Greek society for the similarity between Lydian games and Greek games; that the source would have been aware of occurrences of drought and famine in the geographic region that the Lydians inhabited; and that the source realized that drought and famine in Lydia had the ability to prompt a large migration of Lydians to what is now Italy and thereby had the ability to influence social life in Greece).

[5]ARTHUR KOESTLER, THE GHOST IN THE MACHINE 178 (1967).

functions in a society—and looks at a possible step that a government might take to curb growth in the size of the human population.

1.1 The Neglect of the Population Factor by Environmentalism and Environmental Law Scholarship

In spite of the evidence that human overpopulation has had and will continue to have an array of deleterious effects (*infra* Part 1.4), the negative influence of mounting human numbers on ecosystem functioning is today almost unnoticed by environmentalists[6], a group that pays close attention to the ecosystem of the planet. Regrettably, an important segment of this group—scholars in environmental law—have evidently put on the same blinders[7] even though scholars are supposed to be on guard against error in the assumptions that they and others make. The situation in environmental law is especially curious in the United States where the federal National Environmental Policy Act,[8] which is widely considered a cornerstone of the domestic campaign to protect the environment,[9] explicitly names population growth as a source of environmental harm. Title I of the Act, to be precise, refers to "the profound influences of

[6]John Harte & Robert Socolow, *Impatient Earth, in* EARTH 2020, at 13, 14–15 (Philippe Tortell ed., 2020) (observing that, in 1970, concern with environmental degradation was "deeply intertwined" with concern with population growth but that, since 1970, " 'environment' has distanced itself from 'population' in most discourse"). *Accord*, J. Joseph Speidel et al., *Population Policies, Programmes and the Environment*, 364 PHIL. TRANSACTIONS OF THE ROYAL SOC'Y B 3049, 3049 (2009) (remarking that "the environmental movement's agenda" no longer includes the issue of population growth); *The Debate: Population and the Environment: How Do Law and Policy Respond?*, ENVTL. FORUM, March/April 2017, at 44 (observing that the "environmental movement" has become disconnected from concern with population growth); LEON KOLANKIEWICZ & ROY BECK, FORSAKING FUNDAMENTALS: THE ENVIRONMENTAL ESTABLISHMENT ABANDONS U.S. POPULATION STABILIZATION 56 (Ctr. Immigr. Stud., 2001) (commenting that "[h]istorians need to explain how an environmental issue as fundamental as U.S. population growth could have moved from center-stage within the American environmental movement [during the early 1970s] to virtual obscurity in just 20 years").

[7]Richard Oliver Brooks, *The Gulf Oil Spill: The Road Not Taken*, 74 ALB. L. REV. 489, 512 n.124 (2010–2011); Robert V. Percival, *Looking Backward, Looking Forward: The Next 40 Years of Environmental Law*, 43 ENVTL. L. REP. 10492, 10493 (2013); John K. Setear, *Learning to Live with Losing: International Environmental Law in the New Millennium*, 20 VA. ENVTL. L.J. 139, 163–64 (2001). *See* LOUIS J. KOTZÉ, GLOBAL ENVIRONMENTAL CONSTITUTIONALISM IN THE ANTHROPOCENE 1–2, 4, 7–8 (2016) (listing human population growth as a principal cause of damage to the ecosystem of the Earth; and observing that constitutional law has not expressly recognized the harm to the planetary ecosystem that stems from the human species and, by implication, from numerical growth in the number of human beings).

[8]National Environmental Policy Act, Pub. L. No. 91-190, 83 Stat. 852 (1970) (codified as amended at 42 U.S.C. §§ 4321–4347 (2012)).

[9]*See, e.g.*, Lucia A. Silecchia, *Environmental Ethics from the Perspectives of NEPA and Catholic Social Teaching: Ecological Guidance for the 21st Century*, 28 WM. & MARY ENVTL. L. & POL'Y REV. 659, 770 (2004) (labelling the Act "a landmark in American environmental legislation").

population growth [and] high-density urbanization" on the environment.[10] Notably, the Act has been judicially construed to encompass not only the biosphere but also the overall quality of life in the United States;[11] in the words of one U.S. Court of Appeals, the aim of the Act "is as broad as the mind can conceive."[12] The current failure of U.S. environmental law scholars to address continued population growth is thus baffling even if this failure simply manifests an unawareness among Americans generally of the negative consequences that such growth can have for the ecosystem of the planet.

The inattention that exists among environmentalists and scholars in environmental law to population size and growth is especially puzzling because these groups rely heavily on the concept of an "ecosystem".[13] However, while the ecosystem concept is a key implement in the tool kit of environmentalists, it is, remarkably, not often explicit in environmentally focused law and government policy.[14] The express avoidance of the concept in law and policy may be telling. The Earth is essentially a single ecosystem[15] that, by virtue of the character of a system,[16] is affected by its components and their interactions. One of these components, inescapably, is the human species.[17] The number of people who inhabit Earth, therefore, ought to be of concern to environmentalists and scholars in environmental law. That it is not invites inquiry.

Obviously, the present disregard of the population factor by environmental scholars is not without roots, and these roots ought to be identified because if the roots are known, we may be able to circumvent or eliminate them. In reaching this end, however, a temptation may exist to compare environmental scholars to the proverbial

---

[10]Pub. L. No. 91-190, § 101(a) (1970), 83 Stat. 852 (codified at 42 U.S.C. § 4331(a) (2012)). As seen in *infra* Part 1.2 and Figure 1.2, sizeable increments have occurred in the number of people in the United States since the Act was adopted in 1970, and the increments presently give no sign of materially abating. Similarly, the scale of densely populated places has expanded: The share of the U.S. population that lived in a metropolitan area having at least one million residents rose from 43.7% in 1970 to 59.8% in 2010, an increase of more than one-third. Campbell Gibson, *Chapter 3. Metropolitan and Nonmetropolitan Population, and Large Metropolitan Areas* fig. 3.4, *in* AMERICAN DEMOGRAPHIC HISTORY CHARTBOOK: 1970 TO 2010 (2018) (see section of figure 3.4 labeled "cumulative percent in size category and larger").

[11]Hanly v. Mitchell, 460 F.2d 640, 647 (2d Cir. 1972). *Accord*, Nucleus of Chicago Homeowners Ass'n v. Lynn, 525 F.2d 225, 229 (7th Cir. 1975).

[12]First Nat'l Bank v. Richardson, 484 F.2d 1369, 1377 (7th Cir. 1973).

[13]"[A] dynamic complex of plant, animal and micro-organism communities and their non-living environment interacting as a functional unit." CONVENTION ON BIOLOGICAL DIVERSITY, CBD ARTICLE 2: USE OF TERMS (definition of "ecosystem"), https://www.cbd.int/kb/record/article/6872?RecordType=article (last visited March 19, 2021).

[14]Robert B. Keiter, *Ecosystems and the Law: Toward an Integrated Approach*, 8 ECOLOGICAL APPLICATIONS 332, 333 (1998).

[15]*E.g.*, U.K. Nat'l Ecosystem Assessment, Ecosystems (noting that "[i]n one sense, the entire biosphere of planet Earth is an ecosystem since all its elements interact"), http://uknea.unep-wcmc.org/EcosystemAssessmentConcepts/tabid/98/Default.aspx (last visited March 19, 2021).

[16]17 OXFORD ENGLISH DICTIONARY 496 (2nd ed. 1989) (def. 4 of the word "system").

[17]Jan G. Laitos & Lauren Joseph Wolongevicz, *Why Environmental Laws Fail*, 39 WM. & MARY ENVTL. L. & POL'Y REV. 1, 4–5 (2014).

ostrich.[18] Although doing so may help call attention to the neglect of the population factor, a comparison of environmental scholars to a symbol does not *explain* the blind eye that these scholars have turned to overpopulation. What must be explained is the flawed mental picture of the world on which environmental scholars are evidently relying. Why they have this mental map will tell us why they are inattentive to an agent that is a major source of harm to the ecosystem.

While a complete list of the reasons that environmental scholars in the United States today leave out the population factor is probably impossible to compile, one plausible reason is found in tradition, i.e., social inertia. Americans have traditionally welcomed population growth,[19] at least growth that occurs through childbearing.[20] Therefore, if Americans acknowledged that population increments are harmful, they would experience cognitive dissonance, a psychological state in humans that arises from personal beliefs that are incompatible with one another, with the behavior of the individuals who hold the beliefs, and/or with beliefs endorsed by other individuals.[21] Cognitive dissonance causes individual-level psychological discomfort that is ameliorated by, inter alia, group-effected uniformity in beliefs, including uniformity

---

[18]"One who tries to avoid disagreeable situations by refusing to face them." AMERICAN HERITAGE DICTIONARY 1282 (3d ed. 1992) (def. 2 of "ostrich"). Behavior in humans like that ascribed to the ostrich bird has been the subject of empirical research. *E.g.*, Kaitlin Woolley & Jane L. Risen, *Closing Your Eyes to Follow Your Heart: Avoiding Information to Protect a Strong Intuitive Preference*, 114 J. PERSONALITY & SOC. PSYCHOL. 230 (2018).

[19]Text accompanying *supra* note 1.

[20]*See* Lydia Saad, Gallup, Inc., *Americans, in Theory, Think Larger Families are Ideal* (2020) (reporting the results of national sample surveys of U.S. adults conducted between 1938 and 2018; finding that, during this time, no fewer than roughly one-third, and as many as three-fourths, of all respondents believed that at least three children is the ideal family size), https://news.gallup.com/poll/236696/americans-theory-think-larger-families-ideal.aspx (last visited Sept. 29, 2020). In regions of the world that are socioeconomically developed, a long-term mean of 2.1 births per woman is "replacement-level fertility," and stabilizes the size of the population (assuming, inter alia, that the net immigration rate in these regions is zero). Thomas J. Espenshade et al., *The Surprising Global Variation in Replacement Fertility*, 22 POPULATION RES. & POL'Y REV. 575, 580 tbl. 1 (2004). *See generally* Part 3.4 in *infra* Chap. 3.

Sample surveys conducted during and after the 1950s have found that at least seven out of ten Americans endorse a cap on immigration. Elizabeth Fussell, *Warmth of the Welcome: Attitudes Towards Immigrants and Immigration Policy in the United States*, 40 ANN. REV. SOCIOL. 479, 481, 482 fig. 1 (2014).

[21]Kristen Renwick Monroe et al., *The Psychological Foundations of Identity Politics*, 3 ANN. REV. POL. SCI. 419, 425–26 (2000); David C. Matz & Wendy Wood, *Cognitive Dissonance in Groups: The Consequences of Disagreement*, 88 J. PERSONALITY & SOC. PSYCHOL. 22, 23 (2005).

from the acceptance of tradition.[22] Since the environmental consequences of population growth are inconsistent with social thinking that has long been popular in the United States, they are not easily admitted. As a result, the pernicious influence that population overgrowth has on human societies is not readily accepted or, if accepted at a particular time, is prone to be forgotten.

Tradition, then, is a plausible reason that environmentalists generally do not pay even lip service to modern-day pressures on nature from population size, and instead concentrate on threats to the ecosystem that do not involve the number of people, threats such as consumption practices and industrial technologies. However, tradition is unlikely to be the sole reason. Notably, the population factor was recognized by environmentalists a half-century ago.[23] The obvious question, therefore, is why the link between concern with the environment and concern with overpopulation unraveled, or at least materially weakened. The answer may supply another important reason for the present disregard of population pressures.

In this respect, a credible hypothesis is that general social philosophy in the United States has changed, a hypothesis that may be tested because shifts in law embody shifts in social values.[24] Relevantly, since the beginning of the last third of the twentieth century, law in the United States has moved away from an emphasis on the social responsibilities of the individual and moved toward an emphasis on the personal rights of the individual.[25] The foregoing shift in law is consistent with time-series statistical evidence on social values.[26] Given that Americans have become more self-focused and less committed to the commonweal, they have been making decisions increasingly on the basis of their immediate personal interests, not on the basis of the long-term interests of their society. If U.S. environmentalists recognized this change in social philosophy, as they likely did, they would be uncomfortable now probing matters that are clearly tied to decisions by individuals regarding fertility and

---

[22]Matz & Wood, *supra* note 21, at 35 (concluding that participants in a group exhibit consensus in their beliefs because of pressure from the group); Christopher T. Burris et al., *"By Faith Alone": Religious Agitation and Cognitive Dissonance*, 19 BASIC & APPLIED SOC. PSYCHOL. 17, 28 (1997) (finding that situational discomfort stemming from dissonance involving religious beliefs was curbed by the acceptance of religious tenets). *Cf.* Michael H. Connors & Peter W. Halligan, *A Cognitive Account of Belief: A Tentative Road Map*, FRONTIERS IN PSYCHOL., Feb. 2015, art. 1588, at 1, 4 (summarizing evidence for the proposition that the beliefs of a given individual tend to be mutually affirming). *See generally* Michele F. Margolis, *Cognitive Dissonance, Elections, and Religion: How Partisanship and the Political Landscape Shape Religious Behaviors*, 80 PUB. OPINION Q. 717, 718–19, 734–35 (2016) (finding that the religiousness of Americans is higher when their political-party affiliation does not match the political-party affiliation of the current U.S. President than when it does).

[23]See *supra* note 6.

[24]LARRY D. BARNETT, EXPLAINING LAW: MACROSOCIOLOGICAL THEORY AND EMPIRICAL EVIDENCE 8–9, 51–52 (2015) [hereinafter EXPLAINING LAW].

[25]Elisabeth Zoller, *Citizenship After the Conservative Movement*, 20 IND. J. GLOBAL LEG. STUD. 279, 283–84 (2013). *Cf.* EXPLAINING LAW, *supra* note 24, at 253–61 (providing, for the United States during and after the last one-third of the twentieth century, quantitative measures of change in the prevalence of self-centeredness in social values).

[26]EXPLAINING LAW, *supra* note 24, at 273–83. *See* LARRY D. BARNETT, SOCIETAL AGENTS IN LAW: A MACROSOCIOLOGICAL APPROACH 161–67 (2019) (eras in U.S. constitutional law).

migration, i.e., decisions that determine how many human beings are in a geographic area. Decisions on childbearing and on migration would be regarded as belonging to the individual and an inappropriate topic for everyone else.

1.2 Measures of Human Population Growth

If the thesis is correct that *Homo sapiens* injures the ecosystem through the size and growth of its population, we need to know the magnitude of these demographic agents. Indeed, the thesis is likely to be strengthened by an appreciation of the scale of the human presence on Earth. Part 1.2 provides two graphs that portray human-population size and growth over a lengthy but recent period, and that do so in a manner that I hope will be meaningful as well as intuitively understandable. The graphs are similar to one another in their general structure. In both graphs, the horizontal axis shows calendar years from 1920 onward. Each graph has two vertical axes. The left vertical axis shows the number of years that would be required for the population of the geographic area covered by the graph to increase by a specified number of people, and this axis serves as the reference for the markers connected by short dashes in the inner portion of the graph. At the opposite side, the right vertical axis shows the total number of people inhabiting the covered geographic area. The right vertical axis is the reference for the markers connected by long dashes in the inner portion of the graph.

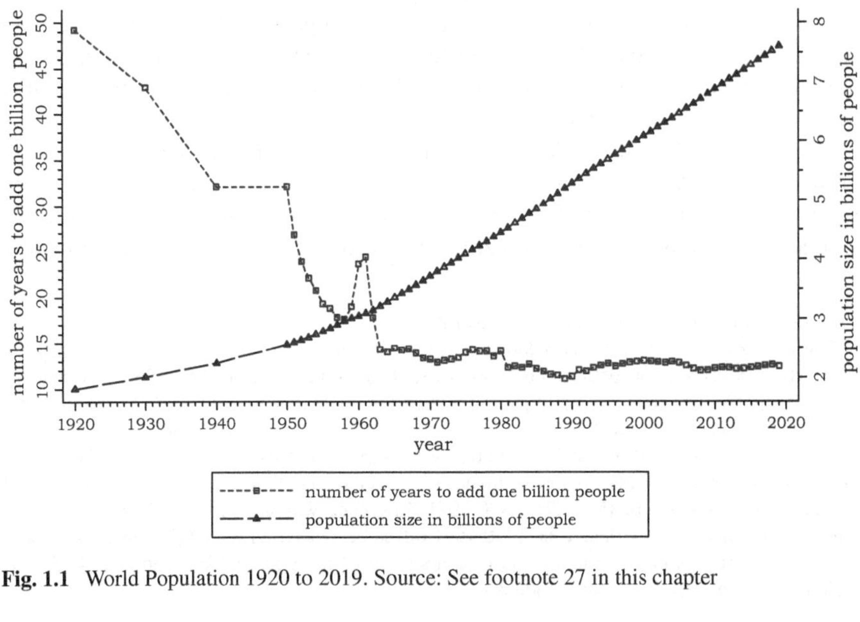

Fig. 1.1 World Population 1920 to 2019. Source: See footnote 27 in this chapter

Fig. 1.2 United States Population 1920–2018. Source: See footnote 32 in this chapter

The two graphs, which are labelled Figs. 1.1 and 1.2, help in understanding the increased environmental impact of *Homo sapiens*. In quantitative terms, the aggregate impact of human activity at a given point in time is the mathematical product of (i) per capita impact, i.e., the average impact of each person in the population, and (ii) the total number of people in the population. Of course, both (i) and (ii) can vary over time, and one of them may be rising while the other is falling. Assuming there is no change in relevant technology and consumption patterns, (i) remains constant, and the graphs portray the heightening environmental impact of the human species that stems from growth in the size of the human population. Otherwise expressed, a focus on (ii) offers insight into the consequences for nature that are due to the absolute number of people and to change in that number.

Figure 1.1 deals with world population and uses data for the period from 1920 through 2019.[27] As seen in the vertical axes, the figure has two yardsticks—(a) the number of years that would be required for one billion people to be added to the population of the planet (left vertical axis) and (b) the total size of the population

[27]The plots in Figure 1.1 were constructed as follows:

Estimates of pre-1950 world population size were obtained for the years 1920, 1930, and 1940 from UNITED NATIONS, DEMOGRAPHIC YEARBOOK: 1960, at 118 tbl. 2 (12th ed. 1960) (estimates of midyear population), *available at* https://unstats.un.org/unsd/demographic-social/products/dyb/#statistics (last visited March 20, 2021). Within the intervals between 1920 and 1930, 1930 and 1940, and 1940 and 1950, linear interpolation was employed to estimate world population size for individual years. The interpolated estimates allowed the calculation of year-to-year numerical differences in world population. The 1920-1921 difference was used for 1920, the 1930-1931 difference was used for 1930, and the 1940-1941 difference was used for 1940. These differences were the basis for computing the number of years required in 1920, in 1930, and in 1940 for the world population to grow by one billion people.

of the world in billions (right vertical axis). To obtain yardstick (a), one billion was divided by the increment that took place in world population size during each year. The number obtained from this division is, therefore, an extrapolation, and a decline in the number evidences a shortening of the period necessary for a one-billion addition to world population.

World population growth, not surprisingly, can be portrayed in other ways. The yardsticks that are used in Figure 1.1 have intuitive appeal because they show both the length of time necessary for the size of the human population to grow by a very large number and the absolute size of the human population at each point in time. Although the trend in annual rates of population growth could be used as a yardstick, a downward trend in the rates will be misleading because, all else being equal, numerical additions to the population increase the impact of humans on their environment.[28] In a given year, a numerical addition to (or reduction in) population size results from how many people were in the population at the start of the year multiplied by the rate of change in this number during the year; when the size of the existing population is large, an above-zero rate of change (even though lower than in the prior year) produces a substantial population increment. Indeed, world population size has been expanding rapidly and is projected to continue doing so.[29] The temporal course of annual rates of change, therefore, is insufficient to grasp the magnitude of the mounting environmental impact of *Homo sapiens*.

As Figure 1.1 reveals, the number of years that would be required for world population to grow by one billion people diminished markedly between 1920 and the late 1950s and then underwent a brief increase followed by a brief decrease. From and after the middle of the 1960s, the billion-people measure has been relatively steady and generally in the range of twelve years through fourteen years. Accordingly, a very large numerical addition to the size of the human population took much longer in the early twentieth century than it takes in the early twenty-first century (49.3 years in 1920 compared to only 12.5 years in 2019). Not surprisingly, since the time necessary for the addition of a billion people has become remarkably short, the presence of *Homo sapiens* on the planet has been expanding rapidly. An illustration of this expansion is found in the data that are the basis of the figure. Specifically, increments of one billion people occurred in world population size during the temporal intervals shown below:

Estimates of midyear world population size for each year from 1950 onward were obtained from U.S. Census Bureau, International Data Base, https://www.census.gov/programs-surveys/international-programs/about/idb.html (data obtained on March 14, 2020). The International Data Base furnished estimates that had been updated in December 2019. The differences in world population between adjacent years were computed and used to calculate, for each year, the number of years that would be required for the population of the world to grow by one billion people.

[28] After the middle of the twentieth century, the yearly rate of world population growth is estimated to have risen to approximately two percent during the period 1970–1975 and subsequently to have undergone a secular decline to 1.18% during the period 2010–2015. 1 UNITED NATIONS POPULATION DIVISION, WORLD POPULATION PROSPECTS: 2019. COMPREHENSIVE TABLES, at 12 tbl. A.5 (2019), https://population.un.org/wpp/Publications.

[29] *Id.* at 2-3 tbl. A.1.

Growth in population size	Interval	Length of interval
from two billion to three billion	1930–1959	29 years
from three billion to four billion	1959–1974	15 years
from four billion to five billion	1974–1987	13 years
from five billion to six billion	1987–1999	12 years
from six billion to seven billion	1999–2012	13 years

Let us now turn to Figure 1.2. Unlike Figure 1.1, which deals with the world as a whole, Figure 1.2 focuses on just the United States. Population growth and size in this one country is important because the consumption of mineral and botanical resources is higher per person among U.S. residents than among the residents of most other nations in the world.[30] An American who bears or sires a child is accordingly responsible for a long-term increase in carbon-dioxide emissions from fossil fuels that is much greater in amount than an individual who bears or sires a child in any of the other ten most highly populated countries.[31] Demographic change in the United States is thus critical to understanding environmental degradation worldwide.

Figures 1.1 and 1.2 differ not only in their geographic focus but also in their measures of population growth and population size. In terms of population growth, Figure 1.1 uses increments of one billion while Figure 1.2 uses increments of nineteen million. In terms of population size, Figure 1.1 is scaled in billions of people while Figure 1.2 is scaled in millions of people. Figure 1.2 thus graphs, for the period 1920 to 2018, (i) the number of years that would be needed for the population of the United States to grow by nineteen million people and (ii) the size of the U.S. population in millions.[32] I chose nineteen million for (i) because nineteen million was roughly the

[30]Sylvia Gierlinger & Fridolin Krausmann, *The Physical Economy of the United States of America: Extraction, Trade, and Consumption of Materials from 1870 to 2005*, 16 J. INDUS. ECOLOGY 365, 373–74 (2011); PAUL HARRISON & FRED PEARCE, *Population and Natural Resources*, *in* AAAS ATLAS OF POPULATION & ENVIRONMENT 43, 44–45 (2000). *See also* Gregory M. Fulkerson et al., *Global Warmers and Global Coolers: A Cross-National Examination of Global Warming Dynamics*, 40 INT'L J. SOCIOL. 44, 48–49, 60–64 app. tbl. 1 (2010) (measuring net impact on climate warming by the population of 135 nations; finding that only two of the 135 nations added more per person to climate warming than the United States). The relatively high level of resource use in the United States was undergirded during the last two or three decades of the twentieth century by an increase in the prevalence among Americans of preferences for personal consumption and enjoyment. EXPLAINING LAW, *supra* note 24, at 253–69.

[31]Paul A. Murtaugh & Michael G. Schlax, *Reproduction and the Carbon Legacies of Individuals*, 19 GLOBAL ENVTL. CHANGE 14, 18 (2009). *See also* Seth Wynes & Kimberly A. Nicholas, *The Climate Mitigation Gap: Education and Government Recommendations Miss the Most Effective Individual Actions*, 12(7) ENVTL. RES. LETTERS, July 2017, at 1, 3, 4 fig. 1, 6 tbl. 1, https://iopscience.iop.org/journal/1748-9326.

[32]The plots in Figure 1.2 are based on population estimates for July 1 of each year. The estimates were obtained from the following sources: U.S. Census Bureau, Historical National Population Estimates, 1900 to 1999 (2000) (data for the resident population except in 1940-1979, when the data are for the total population, i.e., residents plus military personnel stationed outside the U.S.), https://www.census.gov/data/tables/time-series/demo/popest/pre-1980-national.html; U.S. Census Bureau, Intercensal Estimates of the Resident Population by Sex and Age for the United States:

population of the state of New York at the start of the twenty-first century.[33] New York is a logical choice since it is not only populous but socially and economically prominent nationally[34] as well as internationally.[35] Figure 1.2, in short, tells us how many years would have had to elapse at any particular time point during the covered period for there to be a nation-level population increase that was numerically equal to the New York state population when the twenty-first century began. It also tells us how many people were in the United States at a particular time during the covered period.

As is readily apparent in Fig. 1.2, the time interval necessary for a nineteen-million increment in U.S. population size was longest during the 1930s. For almost all of that decade, the number of years needed to add nineteen million people to the population of the nation was either quite close to or well above twenty. However, from the late 1940s onward, less than ten years (and usually less than nine years) were required for the country to grow by nineteen million people. Unsurprisingly, the population of the United States underwent a large absolute increase over the course of this period. Between 1950 and 2016, for example, the U.S. population went from 152 million to 323 million, an addition of *nine* nineteen-million increments. Put differently, in just two-thirds of a century, the United States experienced a population increase equivalent to nine New York state populations around the year 2000.

1.3 Demographic Components of Population Change in the United States

Part 1.3 delves further into the increase in the size of the U.S. population because, as discussed above, Americans have a disproportionately large impact on the environment. An intuitively meaningful quantitative gauge of this impact is the number of Earths that would be needed to maintain the entire population of the world at the per

---

April 1, 2000 to July 1, 2010 (2011), https://www.census.gov/data/tables/time-series/demo/popest/intercensal-2000-2010-national.html; U.S. Census Bureau, Table 1: Annual Estimates of the Resident Population for the United States, Regions, States, and Puerto Rico, April 1, 2010 to July 1, 2018 (2018) (updated Dec. 2018), https://www.census.gov/newsroom/press-kits/2018/pop-estimates-national-state.html (last visited March 15, 2020). Adjacent-year changes in population size were calculated from the estimates of the number of people in the U.S. population each year and were used to determine the number of years necessary for the U.S. population to add nineteen million people.

[33] U.S. CENSUS BUREAU, NEW YORK: 2010, at III-1, 6 tbl. 4 (2012) (reporting that the population of New York state was 18,976,821 on April 1, 2000 and 19,378,102 on April 1, 2010), *available at* https://www.census.gov/prod/www/decennial.html (under "Census of Population and Housing, 2010," open "CPH-2. Population and Housing Unit Counts" and then select "New York" hyperlink). New York, accordingly, had just a small net change in the size of its population during the first decade of the twenty-first century.

[34] RICHARD FLORIDA, CITIES AND THE CREATIVE CLASS 118–120, 122–23 (2005).

[35] Peter J. Taylor, *Leading World Cities: Empirical Evaluations of Urban Nodes in Multiple Networks*, 42 URB. STUD. 1593, 1605–06 & tbl. 12 (2005).

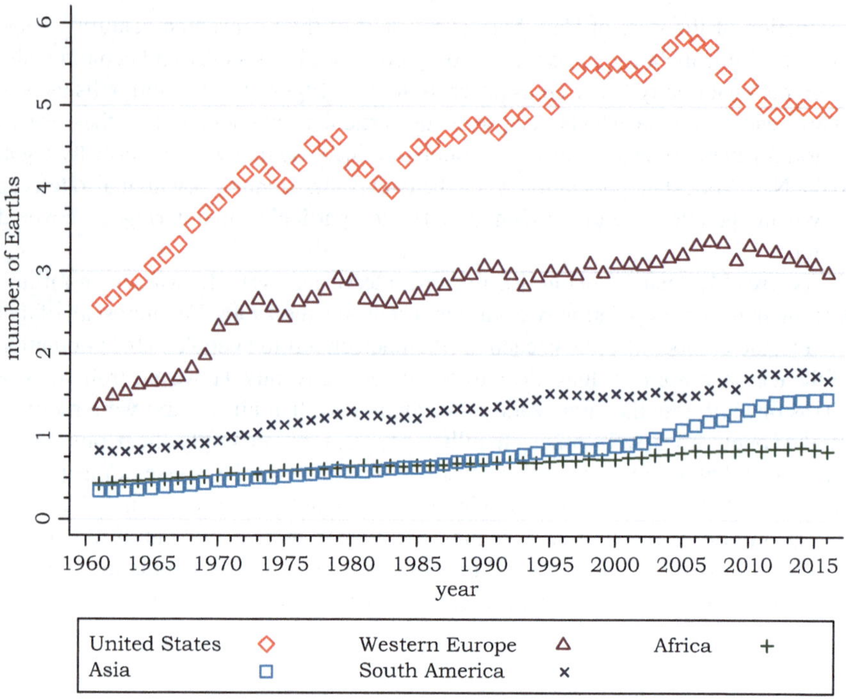

Fig. 1.3 Number of Earths Necessary for the Population of the World to Consume/Dispose Per Capita at the Level of a Given Nation or Region 1961–2016. Source: See footnote 36 in this chapter

capita level of resource consumption and disposal that existed in the United States at a particular point in time. The number of Earths for the United States can then be compared to the number of Earths that would be needed to provide the population of the planet with the per capita resource consumption/disposal level of another country or region at the same point in time. The comparison involves two or more "ecological footprints" and can be made using Fig. 1.3. Estimates of the footprints are available for each year in the period 1961-2016.[36]

[36]The footprints are calculated by the Global Footprint Network (2019), https://data.footprintnetwork.org/#/ (from the "Explore Data" dropdown menu, choose "Reserve/Deficit Trends" hyperlink; on "Country Trends" page, use "Select Country or Region" menu to choose the name of a specific nation or geographic region; and for "Select Type," choose the "Ecological Footprint (Number of Earths)" option). The Network describes the ecological footprint as "[a] measure of how much area of biologically productive land and water an individual, population or activity requires to produce all the resources it consumes and to absorb the waste it generates, using prevailing technology and resource management practices." Global Footprint Network, Glossary (2020), https://www.footprintnetwork.org/resources/glossary.

The ecological footprints that are estimated by the Global Footprint Network are often employed by researchers as their quantitative indicator of the impact of *Homo sapiens* on the ecosystem. This indicator, however, has not escaped criticism. Linus Blomqvist et al., *Does the Shoe Fit? Real versus Imagined Ecological Footprints*, 11(11) PLOS BIOLOGY, Nov. 2013, at 1 (article e1001700)

As is evident in Figure 1.3, the ecological footprint of the United States is much larger than the ecological footprint of South America, Asia, or Africa. While that may not be surprising to readers, the ecological footprint of the United States substantially exceeds the ecological footprint of Western Europe, too. In the most recent year for which the footprints have been published (2016), the population of the world would have needed 5.0 Earths to be able to consume/dispose per capita at the same level as the United States population. To be at the level of per capita consumption/disposal in Western Europe, the population of the world would have needed 3.0 Earths, a number that is high but substantially below the number for the United States. As to the remaining regions, the population of the planet would in 2016 have required 1.7 Earths to be at the per capita consumption/disposal level of the population in South America, 1.5 Earths to be at the per capita consumption/disposal level of the population in Asia, and 0.8 Earths to be at the per capita consumption/disposal level of the population in Africa. To state the preceding differently, the ecological footprint of the United States in 2016 was more than one-and-a-half times larger than the ecological footprint of Western Europe, was about three times larger than the ecological footprint of Asia or South America, and was roughly six times larger than the ecological footprint of Africa. Furthermore, the level of per capita consumption/disposal in the United States has been elevated for decades. Indeed, the number-of-earths measure for the United States has not been below 4.0 since 1983 and has not been below 3.0 since 1964. In simple terms, Americans are living very much beyond their environmental means, and although their footprint is smaller today than at its peak in 2005, it has not shrunk dramatically. The size and growth of the population of the United States therefore matters to the ecosystem of the planet, and matters a lot.

Analyzed demographically, numerical change in the population of a country during a defined period of time stems from (i) the difference between the number of births and the number of deaths in the country (a difference known in demography as "natural" increase or decrease) and (ii) the difference between the number of individuals entering the country (i.e., immigrants) and the number of individuals leaving the country (i.e., emigrants). Since immigration involves lawful immigration as well as unlawful immigration, change in the number of people in a country = (number of births) – (number of deaths) + (number of lawful immigrants) + (number of unlawful immigrants) – (number of emigrants). For a particular country and period of time, then, five demographic phenomena determine whether and by how much the resident population numerically expands or contracts.

In the United States, accurate counts are unfortunately not available for all of the five phenomena. Rather, credible data exist for just three of them. The two phenomena that suffer from deficient data are the number of people who unlawfully enter the United States and the number of people (citizens and noncitizens) who move out of

(describing the Network-estimated footprints as being "so misleading as to preclude their use in any serious science or policy context"). *Contra*, William E. Rees & Mathis Wackernagel, *The Shoe Fits, but the Footprint is Larger than Earth*, 11(11) PLOS BIOLOGY, Nov. 2013, at 1 (article e1001701) (responding to the criticisms advanced by Blomqvist et al. and contending that the Network-estimated footprints are a thorough, and currently the best, measure of nation-level environmental impacts).

the United States.[37] Without reliable counts of unlawful immigration and emigration, the net effect of between-country migration on the number of people in the resident U.S. population cannot be known precisely. It must, instead, be estimated.

Can credible estimates of the net number of international migrants be developed for the United States? The answer is time-bifurcated. In terms of the distant past, a reliable estimate seems beyond reach,[38] but in terms of recent decades, a rough approximation is feasible. My steps in coming up with the approximation—admittedly inexact and perhaps unsatisfactory—are described in the Appendix to this chapter. My conclusion, concisely stated, is that net international migration (lawful and unlawful) was the source of approximately one-fourth of the 1990-to-2010 numerical increase of the resident U.S. population. A share of this magnitude is certainly not trivial. Nor, given the globalized character of the world today, should it be surprising.[39]

The foregoing points need to be kept in mind as we consider Fig. 1.4. The figure, which covers the years 1920 to 2017, deals with annual rates of natural increase in and rates of lawful immigration into the United States over essentially the same period that was the focus of Fig. 1.2.[40] Rates rather than absolute numbers are used in Fig. 1.4 because Fig. 1.4 has twin goals—for each year and across all years from

---

[37] LAURA B. SHRESTHA & ELAYNE J. HEISLER, CONG. RES. SERV., THE CHANGING DEMOGRAPHIC PROFILE OF THE UNITED STATES 11 (2011) (noting that, because of "methodological difficulties," the acquisition of data on emigration from the United States was halted in 1957), https://fas.org/sgp/crs/misc/RL32701.pdf.

[38] In the 1980s, available data on the number of unlawful immigrants in the United States were characterized as being "highly speculative" and as having "gross . . . deficiencies." Vernon Briggs, Jr., *Methods of Analysis of Illegal Immigration into the United States*, 18 INT'L MIGRATION REV. 623, 637 (1984). *Cf.* Mae M. Ngai, *The Strange Career of the Illegal Alien: Immigration Restriction and Deportation Policy in the United States, 1921-1965*, 21 LAW & HIST. REV. 69 (2003) (reviewing the history of the concept of "illegal immigrant" in U.S. immigration law).

[39] The 1990-2010 share should not be extrapolated backward in time and applied to earlier periods. Shares before 1990-2010 may have differed substantially from the share in 1990-2010. Indeed, the share of U.S. population growth due to net international migration may have been much larger during some pre-1990 periods than it was in 1990-2010; in other pre-1990 periods, the share may have been much smaller than in 1990-2010.

[40] Figure 1.4 is based on data obtained from the following sources:

• U.S. CENSUS BUREAU, 2003 STATISTICAL ABSTRACT: MINI- HISTORICAL STATISTICS tbl. HS–8 (yearly number of immigrants lawfully entering the United States per 1,000 U.S. population from 1920 to 2001), tbl. HS-13 (yearly number of births and number of deaths in the United States per 1,000 U.S. population from 1920 to 2001), https://www.census.gov/library/publications/2003/compendia/statab/123ed/hist.html.

• Joyce A. Martin et al., Nat'l Ctr. Health Stat., *Births: Final Data for 2018*, NAT'L VITAL STAT. REP., Nov. 27, 2019, at 12 tbl. 1 (yearly number of births in the United States per 1,000 population, 2010-2018).

• Joyce A. Martin et al., Nat'l Ctr. Health Stat., *Births: Final Data for 2012*, NAT'L VITAL STAT. REP., Dec. 30, 2013, at 15 tbl. 1 (yearly number of births in the United States per 1,000 population, 2002-2009).

• Kenneth D. Kochanek, Nat'l Ctr. Health Stat., *Deaths: Final Data for 2017*, NAT'L VITAL STAT. REP., June 24, 2019, at 21 tbl. 1 (yearly number of deaths in the United States per 1,000 population, 2010-2017).

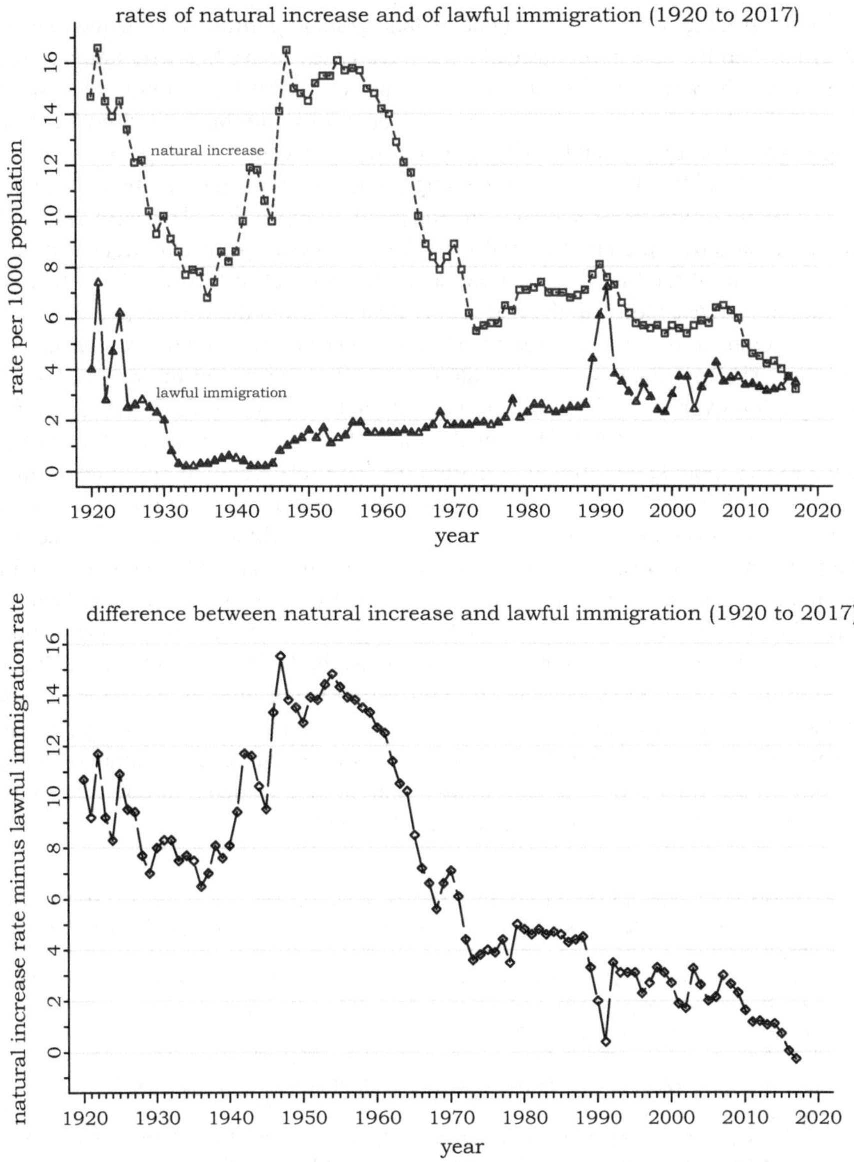

Fig. 1.4 Components of Population Growth in the United States. Source: See footnote 40 in this chapter

1920 to 2017, the figure seeks to render a demographic portrait of the United States in terms of both (i) the importance of natural increase relative to lawful immigration and (ii) the importance of natural increase and of lawful immigration relative to the size of the population of the country. Unlawful immigration is omitted from Fig. 1.4 because I was unable to estimate it reliably except for a single two-decade period (1990–2010). The rate of total immigration, consequently, was higher than the rate of lawful immigration shown in the top graph of Fig. 1.4, and because unlawful immigration added (perhaps materially) to U.S. population growth throughout 1920–2017, the rate of total immigration undoubtedly approached (and reached) the rate of natural increase more rapidly and sooner than shown in the bottom graph.

The graphs in Fig. 1.4 are presented as a pair because they should be considered together. The top graph shows the rate of natural increase and the rate of lawful immigration over the 1920–2017 period. One noteworthy aspect of the top graph is the distance between the lines for the rates. The distance measures the importance to U.S. population growth of natural increase relative to lawful immigration, and it diminishes when these demographic phenomena are becoming more alike in their impact on population size. The bottom graph, which uses a single line for the difference each year between the rate of natural increase and the rate of lawful immigration, makes plain that the rates were virtually identical at the end of the time series in 2017. Natural increase and lawful immigration, in other words, were adding about equally to U.S. population growth in the last half of the second decade of the twenty-first century. However, lawful immigration is not *total* immigration, which includes both unlawful immigration and lawful immigration. Total immigration is necessarily larger in amount than lawful immigration, but since accurate data are lacking on the volume of unlawful immigration into the United States for the entire 1920–2017 period, Fig. 1.4 omits total immigration.[41]

• Jiaquan Xu et al., Nat'l Ctr. Health Stat., *Deaths: Final Data for 2013*, NAT'L VITAL STAT. REP., Feb. 16, 2016, at 19 tbl. 1 (yearly number of deaths in the United States per 1,000 population, 2002-2009).

National Vital Statistics Reports are available at https://www.cdc.gov/nchs/products/nvsr.htm.

I calculated yearly rates of lawful immigration for 2002 to 2017 using data in:

DEP'T OF HOMELAND SECURITY, 2018 YEARBOOK OF IMMIGRATION STATISTICS tbl. 1 (2020) (number of persons granted "lawful permanent resident status," by fiscal year from 1820 to 2018) [hereinafter YEARBOOK OF IMMIGRATION STATISTICS], https://www.dhs.gov/immigration-statistics/yearbook.

U.S. Census Bureau, Table 1: Annual Estimates of the Resident Population for the United States, Regions, States, and Puerto Rico, April 1, 2010 to July 1, 2019, https://www.census.gov/data/tables/time-series/demo/popest/2010s-national-total.html.

U.S. Census Bureau, Table 1. Intercensal Estimates of the Resident Population by Sex and Age for the United States: April 1, 2000 to July 1, 2010, https://www.census.gov/data/tables/time-series/demo/popest/intercensal-2000-2010-national.html.

At the time that I compiled the data for Figure 1.4, final data on U.S. death rates were not available for 2018; as a result, I was unable to calculate the rate of natural increase in 2018. Because the last year in the figure for the rate of natural increase is 2017, the last year in the figure for the rate of immigration is also 2017.

[41] Figure 1.4 seeks to portray the effect of lawful immigration and the effect of natural increase on U.S. population size. Consequently, the figure employs data on the *resident* U.S. population

Why are the preceding points important? One reason is that these points, when joined with the assumption that immigrants remain in their new land and have the same birth and death rates as citizens, allow a significant inference to be drawn from the graphs in Fig. 1.4. Specifically, the graphs indicate that, at the present time, total immigration is probably contributing more than natural increase to growth in the number of people who reside in the United States. Even if the contribution of total immigration just slightly exceeds the contribution of natural increase, total immigration today, when set in the context of the entire 1920–2017 period, is playing a large role in expanding the size of the U.S. resident population.

The magnitude of the impact of immigration on present-day U.S. population growth merits elaboration. The lawful and unlawful flow of people from other nations into the United States has undoubtedly had multiple consequences for American society, some of which consequences were positive and some of which were negative . As an example of a positive consequence, the arts were clearly enriched by the cultures that immigrants brought to America starting around the turn of the twentieth

---

as defined by the U.S. Census Bureau. The Bureau, pursuant to a federal statute that governs the decennial census, deems a U.S. "resident" to be a person who "usually resides in the United States." Act of March 1, 1790, ch. 2, § 5, 1 Stat. 101, 103.

The U.S. Department of Homeland Security ("Department") publishes two time series on the number of people expelled from the United States, but as explained below, each time series includes an unknown number of persons who fall outside the Census Bureau definition of "resident." Because each of these time series counts expelled persons who are not residents, neither of the time series is included in Figure 1.4.

Under the Immigration and Nationality Act , persons who are not U.S. citizens as well as persons ("nationals") who do not "owe[] permanent allegiance to the United States" gain "admission" and are "admitted" into the United States — i.e., become residents of the United States — when they are approved for "lawful entry . . . into the United States after inspection and authorization by an immigration officer." 8 U.S.C. §§ 1101(a)(3), -(13)(A), -21, -(22) (2014). The Department follows these statutory provisions in its two time-series on expulsions from the United States. One of the series consists of data on "removals"; the other time series consists of data on "returns." A "removal" as well as a "return" involves an "alien," i.e., a person who is not a U.S. citizen or a U.S. national. DEP'T OF HOMELAND SECURITY, DEFINITION OF TERMS (2018) (definition of "alien") [hereinafter DEFINITION OF TERMS], https://www.dhs.gov/immigration-statistics/data-standards-and-definitions/definition-terms (last visited March 20, 2021). A "removal" is an expulsion of "an inadmissible or deportable alien" who is subject to "an order of removal"; a "return" is an expulsion of "an inadmissible or deportable alien" who is not subject to "an order of removal." Yearly counts are available for "removals" starting in 1892 and for "returns" starting in 1927. YEARBOOK OF IMMIGRATION STATISTICS, *supra* note 40, at tbl. 39.

Both types of expulsions include "inadmissible" aliens. The Department defines an "inadmissible" alien as "[a]n alien seeking admission at a port of entry who does not meet the criteria in the INA [Immigration and Nationality Act] for admission." DEFINITION OF TERMS, *supra*. Under the foregoing definition, "inadmissible" aliens are not U.S. residents because they do not normally make their home in the United States. Act of March 1, 1790, *supra*. Unfortunately, the Department-supplied counts of yearly "removals" and "returns" include unknown numbers of "inadmissible" aliens and hence unknown numbers of persons who are not U.S. residents. Figure 1.4 cannot utilize the counts, because it requires data on U.S. residents but unspecified numbers of inadmissible aliens are in the time series on removals and returns. On the other hand, "deportable" aliens, who are included in the counts, qualify as U.S. residents because they are "in and admitted to the

century and in key respects passed on to their children.[42] Economic growth in the United States and the personal wealth of Americans benefitted from immigration into the country during 1850-1920, and not just in the short-term—long-term benefits occurred as well.[43] Turning to negative consequences, immigration has been the source of societal strains manifested in political and social divisiveness. For example, a study of immigration into the United States during the period 1910–1930 found that cultural differences between immigrants and natives, not negative economic effects of immigrants on natives, were an important source of public hostility toward immigrants—hostility that was embodied not only in anti-immigrant political ideology and law but also in reduced city-government spending and tax rates.[44] Beyond social friction, and of direct concern to the instant paper, international migration to the United States would have increased pressure on the natural environment of the country.[45]

An appreciation of the preceding consequences, which current immigration also presumably generates, should be accompanied by the recognition that U.S.-bound migrants often leave their home countries because these countries are experiencing

---

United States subject to any grounds of removal specified in the Immigration and Nationality Act." DEFINITION OF TERMS, *supra*.

[42]Charles Hirschman, *Immigration and the American Century*, 42 DEMOGRAPHY 595, 610–13 (2005); Philip Kasinitz, *Immigrants, the Arts, and the "Second Generation Advantage" in New York*, *in* NEW YORK AND AMSTERDAM: IMMIGRATION AND THE NEW URBAN LANDSCAPE 263 (Nancy Foner et al. eds., 2014); Philip Kasinitz, *"Immigrants! We Get the Job Done!": Newcomers Remaking America on Broadway*, 42 ETHNIC & RACIAL STUD. 883 (2019).

[43]Sandra Sequeira et al., *Immigrants and the Making of America*, 87 REV. ECON. STUD. 382, 416 (2020).

[44]Marco Tabellini, *Gifts of the Immigrants, Woes of the Natives: Lessons from the Age of Mass Migration* 3–5, 16–21, 26, 27, 31 (Harvard Business School, Working Paper No. 19-005, 2019). *See* Judith Goldstein & Margaret E. Peters, *Nativism or Economic Threat: Attitudes Toward Immigrants During the Great Recession*, 40 INT'L INTERACTIONS 376, 380–81, 398–99 (2014) (using data from both a panel and repeated cross-sections of respondents in Web surveys conducted during the period 2007-2012; uncovering evidence that negative reactions to immigration become more common as differences increase between the culture of immigrants and the culture of Americans; and finding that, in the context of a sharp downturn in the economy, resistance to immigration stems from personal financial anxiety rather than from personal financial problems). *See also* Tesfaye A. Gebremedhin & Astghik Mavisakalyan, *Immigration and Political Instability*, 66 KYKLOS 317, 323–24, 338 (2013) (using cross-sectional data on 78 nations for the year 2000; finding that the share of foreign-born persons in the population of a nation was inversely related to political stability in the nation; and uncovering evidence that immigrants reduce political stability in a nation when they come from cultures that markedly diverge from the culture of the nation in which they settle). See Daniel Kaufmann et al., *The Worldwide Governance Indicators: Methodology and Analytical Issues*, 3 HAGUE J. RULE OF LAW 220, 223 (2011) (describing the empirical indicator of political instability that constituted the dependent variable in the study by Gebremedhin & Mavisakalyan, *supra*). *Cf.* Thomas C. Wilson, *Americans' Views on Immigration Policy: Testing the Role of Threatened Group Interests*, 44 SOCIOL. PERSP. 485, 487–88, 491 tbl. 3, 493 tbl. 4, 495–96 (2001) (using data from a 1994 survey of a national probability sample of U.S. adults; finding that objections to lawful and unlawful U.S. immigration are more likely when immigrants are believed to be an obstacle to "keep[ing] the country united").

[45]See *infra* Part 1.4. The greater pressure on the environment from immigration may arise indirectly through the gains in personal income that immigrants generate in migrant-receiving nations.

one or more problems that are symptoms of overpopulation[46]—problems such as increased crime,[47] depressed monetary income, and/or inadequate food supplies.[48] A recognition of the role of overpopulation in prompting migration to a wealthy country like the United States is relevant here because it brings us back to the general thesis that an inextricable link exists between the number of human beings in America and the ecosystem. We have, in short, come full-circle.

1.4 Ecological Effects of Population Pressure

My inquiry turns next to the evidence on the negative ecological consequences of human overpopulation and the social problems that they engender. As I attempt to show below, the evidence is extensive, and leads me to be mystified by the unwillingness of environmentalists to face up to the existence of human population pressure

---

According to research on net migration rates in twenty-two nations that are members of the Organization for Economic Co-operation and Development, international immigration has, in these nations, (i) raised the per-person gross domestic product of the working-age population and (ii) lowered the unemployment rate. The economic effects were studied for the years 1987-2009; effect (i) was estimated after adjusting for differences between the twenty-two nations in the purchasing power of their currencies. Ekrame Boubtane et al., *Immigration, Growth, and Unemployment: Panel VAR Evidence from OECD Countries*, 27 LABOUR 399, 404, 409–10, 416 (2013). *Accord*, Ekrame Boubtane et al., *Immigration and Economic Growth in the OECD Countries 1986-2006*, 68 OXFORD ECON. PAPERS 340 (2016).

[46]Mathilde Maurel & Michele Tuccio, *Climate Instability, Urbanisation and International Migration*, 52 J. DEV. STUD. 735, 736, 747 tbl. 5, 749 (2016) (using data from 1960 to 2000 on 222 nations; finding that, in these nations, a larger population and greater climate variability increased urbanization and, in turn, emigration to other nations). *See infra* Part 1.4.

[47]*See* Augustine J. Kposowa et al., *Reassessing the Structural Covariates of Violent and Property Crimes in the USA: A County-level Analysis*, 46 BRIT. J. SOCIOL. 79, 88, 93 tbl. IV, 94 tbl. V (1995) (finding, in data on nearly all counties in the United States, that the rate of violent crime was increased by greater population density as well as by faster population growth). No well-designed studies seem to have investigated whether, in nations other than the United States, the level of population density and the rate of population growth affect the violent-crime rate. However, the relationships in the United States found by Kposowa et al., *supra*, are more likely than not to exist outside the United States. At least in developed nations, areas of high population have a greater prevalence of psychiatric illness than areas of low population, and individuals who suffer from a psychiatric illness have a greater probability of committing a violent crime than individuals who do not suffer from a psychiatric illness (though the statistical association between psychiatric illness and violent crime may appear in only certain historical periods). J. Peen et al., *The Current Status of Urban-Rural Differences in Psychiatric Disorders*, 121 ACTA PSYCHIATRICA SCANDINAVICA 84, 91 (2010); S. Hodgins, *Epidemiological Investigations of the Associations Between Major Mental Health Disorders and Crime: Methodological Limitations and Validity of the Conclusions*, 33 SOC. PSYCHIATRY & PSYCHIATRIC EPIDEMIOLOGY S29, S30–S31, S33, S35 (1998).

[48]In countries where fewer than 2800 calories were available daily per capita in the period 1970-1995, population growth reduced per-person accessibility to calories during the 1980s and 1990s, but not during the 1970s. Øystein Kravdal, *Has Population Growth Restricted Improvements in Food Availability Per Head, 1970-95?*, 55 POPULATION STUD. 105, 107–08, 112 tbl. 2, 115 (2001)

on nature and, in turn, society. The evidence regarding the ecological consequences of human overpopulation is particularly compelling given that the consequences will occur collectively rather than individually.[49] That they will be felt in combination, not singly, is important, because the ability of *Homo sapiens* to manage a particular consequence will probably diminish as more consequences happen at the same time.

What conclusions can be drawn from the evidence? Growth in the size of the human population has, through urbanization, increased average daily ground-level temperatures,[50] caused habitat loss,[51] and produced air pollution.[52] In addition, mounting human population size has resulted in the contamination of rivers, bays,

---

(using data for five-year intervals from 1970 to 1995). The absence of an effect in the 1970s is probably explained by the "Green Revolution." The "Green Revolution," which took place in the 1960s, introduced high-producing cereal crops into world agriculture. Gurdev S. Khush, *Green Revolution: The Way Forward*, 2 NATURE REVIEWS GENETICS 815 (2001). The gains in crop yields starting in the 1960s were presumably observed in the 1970s but were evidently overcome during the 1980s and 1990s by continued population growth.

[49] LAURIE LAYBOURN- LANGTON ET AL., INST. PUB. POL'Y RES., THIS IS A CRISIS: FACING UP TO THE AGE OF ENVIRONMENTAL BREAKDOWN 15–16, 31 (2019), https://www.ippr.org/research/publications/age-of-environmental-breakdown.

[50] Eugenia Kalnay & Ming Cai, *Impact of Urbanization and Land-Use Change on Climate*, 423 NATURE 528 (2003); E. Kalnay & M. Cai, *Impact of Urbanization and Land-Use Change on Climate*, 425 NATURE 102 (2003) (reporting correction). *See* Michael J. Ring et al., *Causes of the Global Warming Observed since the 19th Century*, 2 ATMOSPHERIC & CLIMATE SCI. 401, 412–13 (2012) (finding that natural climate variability was the principal determinant of global trends in temperatures during the first three quarters of the twentieth century but that the scale of human activity was the principal determinant of the global rise in temperatures from the middle of the 1970s to 2010).

[51] Zhifeng Liu et al., *The Relationship between Habitat Loss and Fragmentation during Urbanization: An Empirical Evaluation from 16 World Cities*, 11(4) PLOS ONE, April 28, 2016, at 1, 2, 14, https://journals.plos.org/plosone. The number of animal and plant species in a geographic area is reduced by ecosystem changes, including habitat loss, that are attributable to increases in the size of the human population. *See* Grace E. P. Murphy & Tamara N. Romanuk, *A Meta-Analysis of Declines in Local Species Richness from Human Disturbances*, 4 ECOLOGY & EVOLUTION 91, 92–93, 95 & fig. 2A, 97 fig. 5, 99 (2014) (conducting a meta-analysis of 245 studies; finding that human-induced environmental alterations generally, and habitat loss specifically, had a negative average effect on the number of land-based animal and plant species); text accompanying *infra* note 58.

[52] L. N. Lamsal et al., *Scaling Relationship for NO_2 Pollution and Urban Population Size: A Satellite Perspective*, 47 ENVTL. SCI. & TECH. 7855 (2013).

and lakes,[53] and has enlarged the number of endangered mammal and bird species.[54] The expanding human population has also been posited as the underlying reason for the elimination since 1900 of a vast number of mammal populations worldwide.[55] The impact of human population growth on insects is similarly relevant because insects are critical to a variety of environmental processes on which *Homo sapiens* depends.[56] Long-term studies of insects indicate that, in the geographic areas covered by the studies (primarily North America and Europe), the percentage of insect species in decline is much higher than the percentage of vertebrate species in decline; that roughly one out of three insect species is in danger of becoming extinct; and that one out of ten insect species has already become extinct.[57] The decline in and the disappearance of insect populations is most often attributed by the studies to agents tied to growth in the number of human beings. The agents specifically cited by the studies are mainly (i) the destruction of insect habitats due to the conversion of land from its natural state in order to house, transport, and feed humans, and support the economic activities of humans; and (ii) pollution that involves fertilizers and human-created pesticides applied in agriculture, sewage originating in urban locations, and chemicals used in industry.[58]

---

[53] Nina F. Caraco, *Influence of Human Populations On P Transfers to Aquatic Systems: A Regional Scale Study Using Large Rivers*, *in* PHOSPHORUS IN THE GLOBAL ENVIRONMENT 235, 237, 240 tbl. 2, 244 (Holm Tiessen ed., 1995); Michel Meybeck, *Global Analysis of River Systems: From Earth System Controls to Anthropocene Syndromes*, 358 PHIL. TRANSACTIONS ROYAL SOC'Y 1935 (2003) (synthesis of research findings); F. Oldfield & J. A. Dearing, *The Role of Human Activities in Past Environmental Change*, *in* PALEOCLIMATE, GLOBAL CHANGE AND THE FUTURE 143, 143, 145, 148, 150–51, 156–57, 162 (Keith D. Alverson et al. eds., 2003) (reviewing research on lakes and bays; naming, as underlying drivers of pollution in their water, "population growth" and "population pressure").

[54] Jeffrey McKee et al., *Human Population Density and Growth Validated as Extinction Threats to Mammal and Bird Species*, 41 HUM. ECOLOGY 773, 776 (2013). *See also* INTERGOVERNMENTAL SCIENCE- POLICY PLATFORM ON BIODIVERSITY AND ECOSYSTEM SERVICES, THE GLOBAL ASSESSMENT REPORT ON BIODIVERSITY AND ECOSYSTEM SERVICES: SUMMARY FOR POLICYMAKERS 24 (2019) (pointing out that, among the animal and plant categories that were studied, "the global rate of species extinction is already at least tens to hundreds of times higher than the average rate over the past 10 million years"), https://www.ipbes.net/global-assessment.

[55] Gerardo Ceballos et al., *Biological Annihilation via the Ongoing Sixth Mass Extinction Signaled by Vertebrate Population Losses and Declines*, 114 PROC. NAT'L ACAD. SCI. E6089, E6095 (2017) (studying 177 species of nonhuman mammals on five continents, estimating that "billions of populations" of these mammals have been destroyed during the period covered by the study, and pointing out that population exterminations can presage species exterminations). *See also* Gerardo Ceballos et al., *Vertebrates on the Brink as Indicators of Biological Annihilation and the Sixth Mass Extinction*, PROC. NAT'L ACAD. SCI., June 1, 2020, at 1, 3–5 & fig. 4 (studying forty-eight species of mammals and twenty-nine species of birds; finding that, between the start of the twentieth century and the second decade of the twenty-first century, these species became nearly extinct in substantially more areas of the world).

[56] John E. Losey & Mace Vaughan, *The Economic Value of Ecological Services Provided by Insects*, 56 BIOSCIENCE 311 (2006).

[57] Francisco Sánchez-Bayo & Kris A.G. Wyckhuys, *Worldwide Decline of the Entomofauna: A Review of Its Drivers*, 232 BIOLOGICAL CONSERVATION 8, 9 fig. 1, 16–17 & tbl. 1, 22 (2019).

[58] *Id.* at 19–21 & fig. 5.

More human beings on the planet has had other effects as well. By causing *Homo sapiens* to have more-frequent contact with nonhuman animals, it has facilitated the transmission of viruses to, and increased virus-caused diseases among, humans.[59] The incidence of droughts and the incidence of wildfires are affected, too, by the size of the human population. In terms of droughts, *Homo sapiens* has, by its stepped-up presence and accompanying scale of activities, altered the environment in ways that have reduced precipitation over lengthy periods in large geographic areas.[60] As to wildfires, which have steadily worsened in the United States,[61] an expanding human

[59]C. K. Johnson et al., *Global Shifts in Mammalian Population Trends Reveal Key Predictors of Virus Spillover Risk*, 287 PROC. ROYAL SOC'Y B: BIOLOGICAL SCI. 1, 2, 8–9 (2020), https://royalsocietypublishing.org/doi/10.1098/rspb.2019.2736; David M. Morens et al., *Escaping Pandora's Box — Another Novel Coronavirus*, 382 NEW ENGL. J. MED. 1293 (2020), https://www.nejm.org/doi/full/10.1056/NEJMp2002106. Since 1940, a rise has occurred in infectious diseases among humans; the rise involves a variety of pathogens (e.g., bacteria and viruses) and is traceable to higher human population densities. Kate E. Jones et al., *Global Trends in Emerging Infectious Diseases*, 451 NATURE 990, 991 fig. 1, 992 tbl. 1, [994] (2008). The two principal reasons for the post-1940 rise in these infectious diseases were human-induced changes in land use and increases in agriculture. Elizabeth H. Loh et al., *Targeting Transmission Pathways for Emerging Zoonotic Disease Surveillance and Control*, 15 VECTOR-BORNE & ZOONOTIC DISEASES 432, 434 & fig. 1 (2015). Changes in the use of land and the scale of food-growing activities can be related. MARK S. SMOLINSKI ET AL., MICROBIAL THREATS TO HEALTH: EMERGENCE, DETECTION, AND RESPONSE 75–77 (2003), *available at* https://www.ncbi.nlm.nih.gov/books/NBK221486.

[60]Juan P. Boisier et al., *Anthropogenic and Natural Contributions to the Southeast Pacific Precipitation Decline and Recent Megadrought in Central Chili*, 43 GEOPHYSICAL RES. LETTERS 413, 419 (2016); Dimitris A. Herrera et al., *Exacerbation of the 2013-2016 Pan-Caribbean Drought by Anthropogenic Warming*, 45 GEOPHYSICAL RES. LETTERS 10619, 10622–23 (2018); A. Park Williams et al., *Large Contribution from Anthropogenic Warming to an Emerging North American Megadrought*, 368 SCI. 314 (2020); Xing Yuan et al., *Anthropogenic Shift towards Higher Risk of Flash Drought over China*, 10 NATURE COMM. art. 4661, at 1, 4 (2019).

[61]*E.g.*, MATT LEE-ASHLEY & MICHAEL MADOWITZ, CTR. FOR AMERICAN PROGRESS, TOO HOT TO HANDLE 4–5 (2015) (reporting that, to combat wildfires, the U.S. Forest Service and the U.S. Department of the Interior more than doubled their average yearly spending, measured in terms of the purchasing power of the U.S. dollar in 2014, from $660 million during 1985–1994 to $1.7 billion during 2005–2014; and explaining that this increase occurred because, inter alia, more "homes and structures . . . are being built in wildfire-prone areas"), https://www.americanprogress.org/issues/green/reports/2015/10/22/123801/too-hot-to-handle. Inflation-unadjusted expenditures for firefighting by the U.S. Forest Service and agencies of the U.S. Department of the Interior were $239.9 million in 1985, $1,410.8 million in 2000, and $3,143.3 million in 2018. Nat'l Interagency Fire Ctr., Federal Firefighting Costs (Suppression Only), https://www.nifc.gov/fireInfo/fireInfo_statistics.html (follow hyperlink "Suppression Costs [year-year]") (last visited March 16, 2020). *See also* Patrick Baylis & Judson Boomhower, *Moral Hazard, Wildfires, and the Economic*

population has brought more people into close proximity to forests and thereby increased the frequency of wildfires in areas where people–forest contact occurs.[62]

The extent to which the human species is present, in sum, has had significant deleterious consequences for the ecosystem of the planet. Unfortunately, little is known about the conditions under which and the degree to which these consequences are determined by synergies between the size of the human population and other environmental stressors The ecosystem is, by definition, a system, and in a system, the components of the system may work together to produce an outcome that is not the same, in type and/or in magnitude, as the outcome that would occur if each component acted separately.[63]

The above environmental impacts of human population growth are especially significant because they are taking place in a planetary ecosystem that for millennia experienced relatively little fluctuation in mean temperature.[64] As a result, the impacts cannot be divorced from the human-induced change that is occurring in the climate of the planet. Nor should they be divorced from climate change. In the words of then-President Barack Obama in his 2015 State of the Union address, "no challenge—no challenge—poses a greater threat to future generations than climate change."[65] The ecological impact of changing climate has already had social effects,[66] and these effects are expected to intensify.[67] Not surprisingly, a reciprocal relationship exists climate change and population, and causality runs in both directions: Worldwide,

---

Incidence of Natural Disasters 1, 4 (Stanford Inst. for Econ. Pol'y Research, Working Paper No. 18-044, 2018) (finding that expenditures for combatting wildfires are raised "dramatically" by the initial introduction of private housing into wildlands but that these expenditures are raised little after residential growth exceeds a comparatively low threshold of housing density), https://siepr.stanford.edu/research/publications.

[62]M. L. Chas-Amil et al., *Human-Ignited Wildfire Patterns and Responses to Policy Shifts*, 56 APPLIED GEOGRAPHY 164, 165, 172 (2015). *See also* Jennifer K. Balch et al., *Human-Started Wildfires Expand the Fire Niche across the United States*, 114 PROC. NAT'L ACAD. SCI. 2946, 2948 (2017) (concluding that the human species was mainly responsible for the increases that took place during the 1992-2012 period in the spatial and temporal occurrence of wildfires).

[63]F. Stuart Chapin III et al., *Consequences of Changing Biodiversity*, 405 NATURE 234, 240–41 (2000) (observing that environmental stressors interact nonlinearly and contending that science needs to uncover the factors responsible for such nonlinear interactions). Such a study is Li An et al., *Exploring Complexity in a Human-Environment System: An Agent-Based Spatial Model for Multidisciplinary and Multiscale Integration*, 95 ANNALS ASS'N AM. GEOGRAPHERS 54, 56, 75 (2005) (modeling the impact of human population growth on the area available in China for inhabitation by giant pandas, and finding nonlinear effects).

[64]LAYBOURN- LANGTON ET AL., *supra* note 49, at 11, 14; Will Steffen et al., *Trajectories of the Earth System in the Anthropocene*, 115 PROC. NAT'L ACAD. SCI. 8252 (2018); Johan Rockström et al., *A Safe Operating Space for Humanity*, 461 NATURE 472, 474 (2009).

[65]Barack Obama, Remarks by the President in State of the Union Address (2015), https://obamawhitehouse.archives.gov/the-press-office/2015/01/20/remarks-president-state-union-address-january-20-2015.

[66]Tamma A. Carleton & Solomon M. Hsiang, *Social and Economic Impacts of Climate*, SCI., Sept. 9, 2016, at aad9837–1, –8, –10 tbl. 1 (2016).

[67]U.S. Global Change Research Program, *Ch. 1: Overview*, *in* FOURTH NATIONAL CLIMATE ASSESSMENT: IMPACTS, RISKS, AND ADAPTATION IN THE UNITED STATES (2018) (paragraph

Homo sapiens, by virtue of the numerical size of its population, is through its actions negatively affecting the climate of the planet; at the same time, change in climate is negatively affecting the population of *Homo sapiens*.[68] Each of these relationships (i.e., population-to-climate and climate-to-population) is important.

In terms of the population-to-climate link, growth in the number of human beings would logically have increased the environmental impact of the species, because more humans mean more activities that cause harm to the climate of the planet, including harm through the emission of carbon dioxide into the atmosphere.[69] However, the role of population size in this impact is largely overlooked and/or ignored even though it is not difficult to pinpoint. I begin by noting the magnitude of climate change: The yearly average global (land and ocean) temperature "since 1986 . . . appear[s] to have risen at a more rapid rate than for any similar climatological (20-30 year) time period in at

---

after Figure 1.4) [hereinafter NAT'L CLIMATE ASSESSMENT], https://nca2018.globalchange.gov/chapter/1.

[68]Juliann E. Aukema et al., *Biodiversity Areas under Threat: Overlap of Climate Change and Population Pressures on the World's Biodiversity Priorities*, 12(1) PLOS ONE, Jan. 26, 2017 [no pagination in original] ("Human population growth is both a primary driver of climate change and a key factor in climate change vulnerability and adaptation"), https://journals.plos.org/plosone.

[69]Brian C. O'Neill et al., *Global Demographic Trends and Future Carbon Emissions*, 107 PROC. NAT'L ACAD. SCI. 17521 (2010). *Accord*, Hannes Weber & Jennifer Dabbs Sciubba, *The Effect of Population Growth on the Environment: Evidence from European Regions*, 35 EUR. J. POPULATION 379, 386, 393 tbl. 2 (2019) (analyzing data on 1,033 geographic regions in twenty-two European nations; finding that the regression coefficient for the relationship between the yearly rate of population growth and carbon dioxide emissions was (i) positive, sizeable, and statistically significant in both these regions as a whole and the 822 regions located in Western Europe, but (ii), while positive, was not statistically significant at or below .05 in the 211 regions located in Eastern Europe). *See* U.S. Global Change Research Program, *Ch. 2: Our Changing Climate* ("[m]any lines of evidence demonstrate that human activities . . . are primarily responsible for the climate changes observed in the industrial era, especially over the last six decades") (paragraph after Box 2.2), in NAT'L CLIMATE ASSESSMENT, *supra* note 67, https://nca2018.globalchange.gov/chapter/2; Intergovernmental Panel on Climate Change, *Summary for Policymakers* 6 ("[h]uman activities are estimated to have caused approximately 1.0°C of global warming above pre-industrial levels"), *in* INTERGOVERNMENTAL PANEL ON CLIMATE CHANGE, GLOBAL WARMING OF 1.5° C, at 6 (2018), https://www.ipcc.ch/sr15/chapter/summary-for-policy-makers.

Although the climate was warmed by industrialization, it appears to have been warmed also before industrialization by the activities of a growing human population. Industrialization started in Europe during the initial decades of the nineteenth century. Rondo Cameron, *A New View of European Industrialization*, 38 ECON. HIST. REV. 1, 2–3, 12, 15 fig. 2, 18 fig. 4, 20, 23 (1985); Gy. Cukor, *Some Characteristic Features of Industrialization in Developing, in Advanced Capitalist and in Socialist Countries*, 7 ACTA OECONOMICA 47, 48–49 (1971). Prior to industrialization, European migration to the New World (beginning in the late 1400s) brought indigenous societies into contact with high-mortality diseases that sharply reduced the number of people in these societies. Due to the smaller size of the native population, fires became less frequent and vegetation increased in the Americas; the foregoing changes decreased carbon dioxide in the atmosphere and lowered surface air temperatures globally during the late sixteenth century and early seventeenth century. Alexander

least the last 1,700 years."[70] Because the scale of human action has been shown to go up with the number of human beings,[71] planet-wide warming is empirically traceable to growth in the size of the human population. That warming cannot be blamed on change in the amount of energy from the sun. Although increases and decreases in solar energy produce corresponding increases and decreases in air temperatures on Earth, energy from the sun has been in a downward trend over roughly the last three decades.[72] The downward trend, in the absence of countervailing factors, would lower planetary air temperatures.

We should not forget, of course, that the effects of population size on climate may be entirely offset by innovations in technology and shifts in consumption patterns. Reliance on this possibility may underlie the present disregard of the role of population size in climate change. Unfortunately, there is no assurance that a full offset will occur. At the very least, a full offset to the impact of human population size is more difficult to achieve and less likely to happen as that population grows numerically larger—all else being equal, more natural resources are consumed as more people are added to the population, and more consumption of natural resources increases discharges of pollutants into the atmosphere.[73] Obviously, a total offset has not taken place to date. Had it occurred, there would not have been the post-1986 increase in average global temperature.

Koch et al., *Earth System Impacts of the European Arrival and Great Dying in the Americas after 1492*, 207 QUATERNARY SCI. REV. 13, 20, 21 tbl. 3, 24, 27, 30 (2019). *Cf.* Chiara Uglietti et al., *Widespread Pollution of the South American Atmosphere Predates the Industrial Revolution*, 112 PROC. NAT'L ACAD. SCI. 2349 (2015) (concluding that chemical pollutants in the atmosphere of South America were markedly increased by mining and associated activities on the South American continent more than two hundred years before the continent industrialized).

Pre-industrial climate cooling caused by depopulation implies that the planet was warmed at an earlier time by the activities of a growing population and that the cooling was made possible by the prior warming. Estimates of world population size and growth before (as well as after) industrialization are in Ansley J. Coale, *The History of the Human Population*, SCI. AM., Sept. 1974, at 40, 42–43.

[70] U.S. Global Change Research Program, *Ch. 2: Our Changing Climate* (second paragraph after Key Message 1), in NAT'L CLIMATE ASSESSMENT, *supra* note 67, https://nca2018.globalchange.gov/chapter/2. Global mean temperatures during 2016-2020 and 2011-2020 were the highest ever recorded. WORLD METEOROLOGICAL ORG., STATE OF THE GLOBAL CLIMATE 2020: PROVISIONAL REPORT, at [4] (2020).

[71] DAVID ROSNICK, CTR. FOR ECON. & POL'Y RES., THE CONSEQUENCES OF INCREASED POPULATION GROWTH FOR CLIMATE CHANGE 2, 8–9 & fig. 6 (2014), https://cepr.net/report/the-consequences-of-increased-population-growth-for-climate-change (last visited June 24, 2020).

[72] Jet Propulsion Laboratory, Nat'l Aeronautics & Space Admin., *The Causes of Climate Change* (2020) (see graph "Temperature vs Solar Activity"), https://climate.nasa.gov/causes (last visited Oct. 25, 2020).

[73] Aukema et al., *supra* note 68; O'Neill et al., *supra* note 69; Weber & Sciubba, *supra* note 69. Uncertainty exists, however, regarding the relative magnitude of the impact of each source of change (population size, technology, and consumption patterns) on the degree to which resources are used and the environment is damaged. Thomas Dietz & Eugene A. Rosa, *Rethinking the Environmental Impacts of Population, Affluence and Technology*, 1 HUM. ECOLOGY REV. 277, 281 (1994).

While the population-to-climate relationship has gone mostly unrecognized, the climate-to-population relationship has not. Researchers are studying the ramifications that worldwide climatic change has already had and will in the future have for human societies[74]—ramifications that obviously are magnified by an expanding human population. The work of these researchers has led to the conclusion that, inter alia, hurricanes and their storm equivalents (viz., cyclones and typhoons) have been made more intense by atmospheric warming,[75] though not more frequent globally.[76] Since a hurricane at a particular level on the five-level hurricane scale produces roughly four times more damage than a hurricane at the next lower level,[77] hurricanes have, unsurprisingly, had growing economic effects: Between 1900 and 2018, economic damage in the United States from the hurricanes and cyclones that were the most damaging rose substantially even after adjusting for the increase that occurred during this period in the amount of wealth that was exposed to hurricanes and cyclones.[78] Notably, the demographic effects of a hurricane on the human population do not immediately dissipate when the hurricane is over.[79]

As to the future, climate change in coming decades is expected to entail, inter alia, heat waves that will probably increase the number of human deaths, with greater increases in geographic areas close to the equator than in geographic areas far from

---

[74]See the section labeled "Introduction" in Chapter 1 of NAT'L CLIMATE ASSESSMENT, *supra* note 67.

[75]James P. Kossin et al., *Global Increase in Major Tropical Cyclone Exceedance Probability over the Past Four Decades*, PROC. NAT'L ACAD. SCI., May 18, 2020, at 1 (studying hurricanes and their storm equivalents during the period 1979-2017; finding a statistically significant rise over time in the fraction of these storms that were most intense, i.e., were in hurricane category 3, 4, or 5; and observing that the rise is in line with simulations), https://www.pnas.org/content/early/2020/05/12/1920849117. Hurricane categories are based on "sustained surface wind speed"; a hurricane in category 3 has a wind speed of 111 to 129 miles per hour; hurricane categories 4 and 5 involve wind speeds above 129 miles per hour. Nat'l Hurricane Ctr. & Cent. Pac. Hurricane Ctr., Nat'l Oceanic & Atmospheric Admin., *The Saffir-Simpson Hurricane Wind Scale* [1, 3] (2019) [hereinafter *Saffir-Simpson Hurricane Wind Scale*], https://www.nhc.noaa.gov/aboutsshws.php (select "About the Saffir-Simpson Hurricane Wind Scale" hyperlink).

[76]Hiroyuki Murakami et al., *Detected Climatic Change in Global Distribution of Tropical Cyclones*, 117 PROC. NAT'L ACAD. SCI. 10706, 10711–12 (2020) (finding that, over the course of 1980-2018, climate warming did not have an obvious impact on how often cyclones, hurricanes, and typhoons occurred in the world as a whole but did have region-specific impacts on their frequency). During the 1980-2018 period, the number of cyclones/hurricanes increased both in the region that includes Hawaii and in the North Atlantic region. *Id.* at 10706.

[77]*Saffir-Simpson Hurricane Wind Scale*, *supra* note 75, at [1, 3].

[78]Aslak Grinsted et al., *Normalized US Hurricane Damage Estimates Using Area of Total Destruction, 1900-2018*, 116 PROC. NAT'L ACAD. SCI. 23942, 23945 fig. 2 (2019) (concluding that hurricanes/cyclones causing the greatest damage in the United States during 1900-2018 became more frequent at a centenary rate of 330%).

[79]John R. Logan et al., *Trapped in Place? Segmented Resilience to Hurricanes in the Gulf Coast, 1970-2005*, 53 DEMOGRAPHY 1511, 1516, 1522, 1525 tbl. 2, 1526 fig. 2, 1531 (2016) (studying annual population change over three-year periods in 476 counties located within 200 miles of a coastline in six southern/southeastern U.S. states during 1970-2005; finding that a hurricane reduced the growth rate of the total population for all three years in counties suffering damage from the hurricane and reduced the growth rate of the total population for two years in adjacent counties).

it and with greater increases without effective strategies of mitigation and adaptation than with them.[80] Additionally, climate change will entail higher ground-level temperatures that are likely to result in significantly lower crop yields.[81] An important reason for lower crop yields is that agricultural productivity is damaged by droughts.[82] In this regard, a point meriting emphasis is that since the middle of the twentieth century, droughts stemming from higher air temperatures and reduced precipitation have become more frequent in most regions of the world.[83] The higher air temperatures that accompany climate warming are thus projected to bring about an overall net international increase in droughts that affect land devoted to agriculture.[84] This projection, if correct, will have a major impact on human welfare worldwide—an impact that inescapably involves American agriculture. The United States, which exports more food (in terms of economic value) than any other nation,[85] uses about fifty-three out of every one hundred acres of its total land surface for agriculture

[80]Yuming Guo et al., *Quantifying Excess Deaths Related to Heatwaves under Climate Change Scenarios: A Multicountry Time-Series Modelling Study*, 15(7) PLOS ONE, July 31, 2018 [no pagination in original], https://journals.plos.org/plosone. *See also* Emmanuel A. Odame et al., *Assessing Heat-Related Mortality Risks among Rural Populations: A Systematic Review and Meta-Analysis of Epidemiological Evidence*, 15(8), 1597 INT'L J. ENVTL. RES. & PUB. HEALTH 1, 3, 10, 11, 12 app. A (2018) (conducting a meta-analysis of studies investigating the impact of hot weather on levels of human mortality in rural areas of nations located on different continents; finding that hot weather increased human mortality among rural populations, especially in developing countries; and pointing out that hot weather has usually been found to increase human mortality in urban populations, too).

[81]*E.g.*, Guoyong Leng & Maoyi Huang, *Crop Yield Response to Climate Change Varies With Crop Spatial Distribution Pattern*, SCI. REP., May 7, 2017, at 1, 5; Wolfgram Schlenker & Michael J. Roberts, *Estimating the Impact of Climate Change on Crop Yields: The Importance of Nonlinear Temperature Effects* 32 (Nat'l Bureau of Econ. Research, Working Paper No. 13799, 2008).

[82]Isabel Meza et al., *Global-Scale Drought Risk Assessment for Agricultural Systems*, 20 NAT. HAZARDS & EARTH SYS. SCI. 695, 695–96 (2020); Yusuke Kuwayama et al., *Estimating the Impact of Drought on Agriculture Using the U.S. Drought Monitor*, 101 AM. J. AGRIC. ECON. 193, 208 (2018).

[83]Jonathan Spinoni et al., *A New Global Database of Meteorological Drought Events from 1951 to 2016*, 22 J. HYDROLOGY: REGIONAL STUD. 1, 16–18 (2019).

[84]Aiguo Dai et al., *Climate Change and Drought: A Precipitation and Evaporation Perspective*, 4 CURRENT CLIMATE CHANGE REP. 301, 309 (2018).

[85]U.N. FOOD & AGRICULTURAL ORG., WORLD FOOD AND AGRICULTURE STATISTICAL POCKETBOOK: 2018, at 34 chart 61 (2018) (data for 2006 and 2016 on international trade in agricultural products measured in billions of U.S. dollars).

and purposes closely related to agriculture.[86] Climate change, by destabilizing food production in the United States, can therefore disrupt food stocks in other countries.[87]

Social consequences, too, are anticipated from the higher ground-level temperatures that are and will be associated with climate change. These consequences are likely to include more violent crime and property crime;[88] more "collective violence," i.e., violence perpetrated by groups of people for social, economic, and/or political

[86]Daniel Bigelow, *Chapter 1.2 – Major Land Uses in the United States*, *in* AGRICULTURAL RESOURCES AND ENVIRONMENTAL INDICATORS, 2019, at 7 (Daniel Hellerstein et al. eds., 2019) (Econ. Res. Serv. Info. Bull. No. 208).

The Missouri River Basin has been a major agricultural region for grain and livestock production in the United States — for example, almost half of U.S.-grown wheat and fully a third of U.S.-raised cattle come from the Basin. Vikram M. Mehta et al., *Decadal Climate Information Needs of Stakeholders for Decision Support in Water and Agriculture Production Sectors: A Case Study in the Missouri River Basin*, 5 WEATHER, CLIMATE, & SOC'Y 27, 28–29 (2013). The Missouri River Basin, which is divided into upper and lower portions, extends southward in the United States from Montana and North Dakota to Colorado, Kansas, and Missouri. Connie A. Woodhouse, Missouri River Hydroclimatology Overview (presentation to workshop sponsored by the Western States Water Council and the California Dep't of Water Resources, May 22-24, 2019), *available at* https://www.westernstateswater.org/wswc-cdwr-sub-seasonal-to-seasonal-s2s-precipitation-forecasting-workshop.

A study of stream flow in the Upper Missouri River Basin found that the impact of air temperatures on stream flow was more often positive than negative from 1900 until the mid-1980s but that the impact of warming air temperatures on stream flow was more often negative than positive from the mid-1980s to the end of the study period in 2010; the negative impact on stream flow of warming air temperatures since the mid-1980s, the authors opined, could have been responsible for the drought in the Basin during the first decade of the twenty-first century, a drought that may have been unequaled in severity during the past one thousand years. Justin T. Martin et al., *Increased Drought Severity Tracks Warming in the United States' Largest River Basin*, 117 PROC. NAT'L ACAD. SCI. 11328, 11331, 11332 & fig. 3B (2020).

[87]Jerry Hatfield et al., *Chapter 6: Agriculture*, in CLIMATE CHANGE IMPACTS IN THE UNITED STATES: THE THIRD NATIONAL CLIMATE ASSESSMENT 150, 151, 162 (J. M. Melillo et al. eds., 2014). The disruptions are likely to be especially serious for people in other countries whose daily per capita food intake is below 2100 calories. In seventy-six countries classified as middle-income or below, the number of such people in 2020 is believed to be, and the number of such people in 2030 is expected to be, no less than 400 million (excluding the effects of COVID-19 infections on food production and affordability). FELIX BAQUEDANO ET AL., U.S. DEP'T OF AGRICULTURE, INTERNATIONAL FOOD SECURITY ASSESSMENT, 2020- 30, at 1–2, 60 app. tbl. 1-1 (GFA-31) (2020).

[88]Matthew Ranson, *Crime, Weather, and Climate Change*, 67 J. ENVTL. ECON. & MGMT. 274, 276–77, 287 (2014); Dennis M. Mares & Kenneth W. Moffett, *Climate Change and Crime Revisited: An Exploration of Monthly Temperature Anomalies and UCR Crime Data*, 51 ENV'T & BEHAV. 502,

ends;[89] more geographic movement by people (and attendant social disruptions) due to climate-induced increases in sea level;[90] and more instances of social instability caused by outbreaks of infectious diseases.[91] Even a society that has an extensive health-care institution may experience social instability in future outbreaks of infectious diseases, because the public health agencies in the society may be hampered in dealing with an outbreak of a contagious disease and averting an epidemic of the disease not only by the scale of the outbreak but also by the social philosophies that shape the structure of the public health function in the society. The ability of contagious diseases to destabilize a society may, therefore, be partly affected by the dominant social philosophies in the society. In the United States, the twin social philosophies of federalism and individualism create differences in responsibility for public health between the smallest societal element and the largest societal element. Federalism assigns primary responsibility for public health to the smallest governmental entities (viz., state and local governments) and secondary responsibility for

---

512, 524 (2018). Hotter temperatures have been found to increase suicide rates as well as rates of murder, assault, robbery, burglary, and vehicle theft. Ryo Takahashi, *Climate, Crime, and Suicide: Empirical Evidence from Japan* 1, 4-5, 19 tbl. 2, 20 tbl. 3 (Institute for Advanced Study, Waseda Univ., Discussion Paper No. 2016-005, 2017), https://www.waseda.jp/inst/wias/en/research/publication (scroll down page to "Discussion Papers").

[89]Barry S. Levy et al., *Climate Change and Collective Violence*, 38 ANN. REV. PUB. HEALTH 241, 245, 251 (2017) (reviewing pertinent studies and basing its conclusion on "[t]he weight of research evidence"). *See also* Philip Nel & Marjolein Righarts, *Natural Disasters and the Risk of Violent Civil Conflict*, 52 INT'L STUD. Q. 159, 166–67, 170, 175, 179–180 (2008) (using data on 187 political units worldwide that had at least 150,000 inhabitants during the period from 1950 to 2000; studying whether the number of climate-involved catastrophic events per capita raised the likelihood that one or more instances of domestic armed conflict would occur in a year; finding a delayed if not immediate effect; and concluding that disasters stemming from abrupt changes in climate elevate the probability of civil conflict). *Cf.* Erin Llwyd Owain & Mark Andrew Maslin, *Assessing the Relative Contribution of Economic, Political and Environmental Factors on Past Conflict and the Displacement of People in East Africa*, 4 PALGRAVE COMM. 1, 3, 6–7 & tbl. 3 (2018) (using data on ten nations in East Africa during the period from 1963 to 2014; studying the frequency of episodes of political violence, including war, that resulted in a minimum of 500 fatalities per episode; and concluding that, in these nations, population growth but not drought raised the likelihood of such episodes, although the effect of population growth involved a delay of ten years), *available at* https://www.nature.com/articles/s41599-018-0096-6.

[90]Mathew E. Hauer, *Migration Induced By Sea-Level Rise Could Reshape the US Population Landscape*, 7 NATURE CLIMATE CHANGE 321, 323 (2017) (estimating that, in the United States, a rise of 1.8 meters in sea level may prompt millions of people to move and that this migration could result in net population gains or declines in fully 56% of all U.S. counties).

[91]Sadie J. Ryan et al., *Global Expansion and Redistribution of Aedes-borne Virus Transmission Risk with Climate Change*, 13(3) PLOS NEGLECTED TROPICAL DISEASES, March 28, 2019, at 1, 13–15 (modelling climate-induced geographic shifts in the presence of the *Aedes* mosquito genus; finding that, by the year 2080, infectious diseases caused by *Aedes*-transmitted viruses will generally become more widespread and frequent due to climate change); Shannon M. Fast et al., *Modelling the Propagation of Social Response During a Disease Outbreak*, 12(104) J. OF THE ROYAL SOC'Y INTERFACE, March 2015, at 1, 8 (modelling the occurrence of disruptions in social life due to outbreaks of infectious diseases; finding that social disruptions often happen when an

public health to the largest governmental entity (viz., the federal government).[92] Individualism places the greatest responsibility for public health on the smallest societal units (viz., individual human beings) and the least responsibility for public health on the largest unit (viz., society).[93] Notably, federalism and individualism have been linked to the performance by government in situations that involve a potential or actual public health crisis: The primacy of state and local governments and of individual humans in American society is thought to reduce the ability of the governmental institution in the United States to (i) prevent a contagious disease that is already in the country from evolving into an epidemic and (ii) combat a domestic epidemic of a contagious disease when such an epidemic develops.[94]

I close Part 1.4 with a crucial point. The environmental effects of human population size and growth, as stated earlier, are happening concurrently rather than one at a time. This point warrants emphasis because each co-occurring effect may not arise and function independently but, instead, involve synergies. The effects, we must not forget, are occurring in an environmental *system*, and a system has feedback loops.[95] In terms of natural disasters, for example, "[r]isk drivers and consequences are multiplying and cascading, colliding in unanticipated ways."[96] If synergies take

---

outbreak involves a disease having severe health consequences, but that social disruptions do not often happen when an outbreak involves a disease having relatively mild health consequences even though the disease spreads widely and infects many people); Jeff Collmann et al., *Measuring the Potential for Mass Displacement in Menacing Contexts*, 29 J. REFUGEE STUD. 273, 275, 285, 286 fig. 1, 287 & fig. 2 (2016) (modelling responses by families in Somalia to threat-posing events, including epidemics of disease; finding that, once such an event lasted roughly three months, existing social arrangements deteriorated as manifested in, inter alia, planned and unplanned geographic emigration). *Accord*, Lawrence O. Gostin & Ana S. Ayala, *Global Health Security in an Era of Explosive Pandemic Potential*, 9 J. NAT'L SECURITY L. & POL'Y 53, 53–56, 58 (2017) (pointing out that infectious disease outbreaks are likely to be more common in the future because of, inter alia, (i) shifts in climate that promote the spread of mosquitos and mosquito-carried viruses and (ii) increases in population density stemming from growth in the number of people; observing that recent outbreaks of infectious disease demonstrated that public-health agencies are wanting in their capacity to keep such outbreaks from becoming epidemics; and citing instances of social disruption caused by recent outbreaks). In addition, see *supra* note 59 and its accompanying text.

[92]Polly J. Price, *Sovereignty, Citizenship, and Public Health in the United States*, 17 N.Y.U. J. LEGIS. & PUB. POL'Y 919, 930 (2014).

[93]WENDY E. PARMET, POPULATIONS, PUBLIC HEALTH, AND THE LAW 19, 111 (2009).

[94]*Id.* at 14, 192–93, 198–203; Polly J. Price, *Do State Lines Make Public Health Emergencies Worse? Federal versus State Control of Quarantine*, 67 EMORY L.J. 491, 494, 499–500, 509, 516 (2018).

[95]Elizabeth T. Borer et al., *Elements of Disease in a Changing World: Modelling Feedbacks between Infectious Disease and Ecosystems*, ECOLOGY LETTERS, Oct. 12, 2020, at 1; WORLD METEOROLOGICAL ORG., *supra* note 70, at [23–24].

[96]Mami Mizutori & Debarati Guha-Sapir, *Foreward*, *in* HUMAN COST OF DISASTERS: AN OVERVIEW OF THE LAST 20 YEARS, 2000- 2019, at 3 (Ctr. for Res. on the Epidemiology of Disasters, U.N. Office for Disaster Risk Reduction, [2020]), https://reliefweb.int/report/world/human-cost-disasters-overview-last-20-years-2000-2019. The foregoing report deals with disasters that involve geophysical events such as earthquakes and volcanic eruptions; hydrological events such as floods and ocean waves; meteorological events such as storms and extreme temperatures; and climatological events such as droughts and wildfires. Biological events such as epidemics and

place, each effect will be intensified by another effect (or other effects), and together they will inflict greater damage on the biosphere and on society than if each had occurred alone. The whole, in other words, may prove to be greater than the sum of its parts.

Appendix: Estimation of Net Number of International Migrants to the United States, 1990–2010

To approximate unlawful immigration and emigration during the period from 1990 to 2010, I started with estimates, developed by the Pew Research Center, of the number of non-U.S. citizens who were unlawfully in the United States[97]—3,500,000 in 1990; 8,600,000 in 2000; and 11,400,000 in 2010.[98] The size of the unlawful-immigrant population in a given year, of course, is the result of (i) unlawful immigration into the country that occurred over time and gradually built up and (ii) emigration out of

---

invasions of insects, however, are excluded from the report. A "disaster" is defined as an event in which at least ten people were reported to have died, at least 100 people were reported to have been "affected," a state of emergency was declared, and/or international assistance was requested. *Id.* at 8.

[97]The number of U.S. residents who are unlawfully in the country is generally agreed to be much larger than the number that is reported by decennial censuses. WILLIAM P. O'HARE, DIFFERENTIAL UNDERCOUNTS IN THE U.S. CENSUS: WHO IS MISSED? 79 (2019). However, social science estimates of the number of unlawfully present residents cover a wide range. Robert Warren & John Robert Warren, *Unauthorized Immigration to the United States: Annual Estimates and Components of Change, by State, 1990 to 2010*, 47 INT'L MIGRATION REV. 296, 296–97 (2013) (observing that estimates of demographic phenomena rarely "diverge as widely or receive so much public attention" as estimates of the number of resident and entering unlawful immigrants). *See also* Mohammad M. Fazel-Zarandi et al., *The Number of Undocumented Immigrants in the United States: Estimates Based on Demographic Modeling with Data from 1990 to 2016*, 12(1) PLOS ONE, Sept. 21, 2018, at 1, 8 & fig. 1, 10 & fig. 2 (simulating, for each year from 1990 to 2016, the size of the resident U.S. population that was unlawfully in the country), https://journals.plos.org/plosone.

[98]JEFFREY S. PASSEL & D'VERA COHN, PEW RES. CTR., U.S. UNAUTHORIZED IMMIGRANT TOTAL DIPS TO LOWEST LEVEL IN A DECADE 5 (2018), https://www.pewhispanic.org/2018/11/27/u-s-unauthorized-immigrant-total-dips-to-lowest-level-in-a-decade. Unlawfully present immigrants declined to 10.7 million in 2016, the latest year for which estimates were computed. *Id.* An unknown number of the illegally present individuals in the Pew Research Center estimates entered the country lawfully but did not leave when they no longer satisfied the conditions of their admission. *Id.* at 2 (definition of "unauthorized immigrants"). The numbers of unlawfully present foreign-born persons that are estimated for 1990, for 2000, and for 2010 by Passell & Cohn, *supra*, are the same as or very similar to the numbers estimated by Warren & Warren, *supra* note 97, at 315 tbl. 3. The earliest year for which the U.S. Department of Homeland Security publishes such estimates is 2000; the Department estimate for 2000 and its census-revised estimate for 2010 are close to the numbers estimated by Passell & Cohn. OFFICE OF IMMIGRATION STAT., U.S. DEP'T OF HOMELAND SECURITY, POPULATION ESTIMATES: ILLEGAL ALIEN POPULATION RESIDING IN THE UNITED STATES: JANUARY 2015, at 12 tbl. A2-1 (2018), https://www.dhs.gov/immigration-statistics/population-estimates/unauthorized-resident.

If the estimates by Passell & Cohn are correct, non-U.S. citizens whose presence in the country was illegal were 1.4% of the total U.S. population in 1990, 3.1% of the total U.S. population in 2000,

the country during this time by unlawful immigrants. Because the estimates prepared by the Pew Research Center are for unlawful immigrants who in a given year *reside* in the United States,[99] the estimates have dealt with both (i) and (ii). On the basis of these estimates, total net unlawful immigration into the United States during the ten years from 1990 to 2000 was (8,600,000 – 3,500,000) = 5,100,000, or 510,000 per year. During the next ten years, i.e., 2000 to 2010, total net unlawful immigration into the United States was (11,400,000 – 8,600,000) = 2,800,000, or 280,000 per year. Since the average annual volume of lawful immigration into the United States during the twenty years from 1990 to 2010 was approximately 1,000,000,[100] lawful immigration numerically exceeded net unlawful immigration during the last decade of the twentieth century and the first decade of the twenty-first century. Nonetheless, unlawful immigration made a sizeable contribution to the growth of the overall U.S. population at this time, and because unlawful immigration could have made such a contribution outside the 1990-2010 period, it should be presumed not to have been insignificant at any point in U.S. history.

Lawfully entering immigrants who did not stay in the United States must also be considered: The *net* number of immigrants, not the total number of immigrants, determines the number of people that immigration adds to the population of a country. Estimates of the extent of emigration from the United States by foreign-born individuals cover the period since the 1960s, and most of these estimates have placed the annual emigration rates of foreign-born individuals in the range of 1.0–1.5%.[101] The preceding range, however, is an average that aggregates all lengths of stay in the United States by lawfully present and unlawfully present foreign-born persons. Annual emigration rates decline substantially as residence in the country lengthens;[102] specifically, estimated annual emigration rates are 2.7% in the first five post-immigration years, 1.75% in the second five post-immigration years, 1.24% in the third five post-immigration years, and 0.95% in the fourth five post-immigration

---

and 3.7% of the total U.S. population in 2010. Calculated using data from the decennial censuses of 1990, 2000, and 2010. The data are available in CAMPBELL GIBSON, AMERICAN DEMOGRAPHIC HISTORY CHARTBOOK: 1790 TO 2010, at fig. 1-1 (2012).

[99] PASSEL & COHN, *supra* note 98, at 36–41.

[100] Computed from: YEARBOOK OF IMMIGRATION STATISTICS, *supra* note 40, at tbl. 1. Counts of lawfully admitted immigrants are separate from counts of refugees. The yearly number of refugees was between 111,000 and 123,000 during 1990-1994. Refugees numbered less than 100,000 per year from 1995 to 2010; within this period, the number was usually below 75,000. *Id.* at tbl. 13 (reporting the number of refugees coming to the United States during each fiscal year from 1980 to 2017).

[101] Jonathan A. Schwabish, *Identifying Rates of Emigration in the United States Using Administrative Earnings Records*, INT'L J. POPULATION RES., 2011, at 1, 7, 8 fig. 2.

[102] *Id.* at 10 tbl. 3 (covering foreign-born individuals who earned income in the United States during the period 1978-1998). The rates include all foreign-born individuals in the United States, i.e., those who were, and those who were not, lawfully admitted to the country. Although foreign-born persons who lawfully entered the United States and foreign-born persons who did not may have had (and probably did have) different emigration rates, I have no evidence for a specific difference and hence will assume that the two groups exhibited essentially the same rates of emigration.

years.[103] I applied the foregoing rates in ascertaining the net number of lawful immigrants during the last decade of the twentieth century and the first decade of the twenty-first century who were in the U.S. population in 2010. By adding the net number of unlawful immigrants derived above from the data of the Pew Research Center, I obtained the approximate net number of lawful and unlawful international migrants in this period who resided in the United States, i.e., the number of all international immigrants who lawfully or unlawfully came to the United States over the course of the twenty years from 1990 through 2009 and who were in the country as of 2010.[104] This number was roughly 16,500,000. Since the resident U.S. population grew by approximately 60,000,000 people during the same period,[105] net international lawful and unlawful immigration was responsible for somewhat more than one-fourth ((16,500,000 ÷ 60,000,000) × 100 = 27.5%) of the 1990–2010 increase in the number of U.S. inhabitants.

---

[103] *Id.* at 10 tbl. 3.

[104] I estimated the net number of lawful international migrants using the twenty cohorts of immigrants who lawfully entered the United States in each year from 1990 through 2009. A cohort was composed of immigrants who came to the United States lawfully in a particular year. For the size of each cohort, I employed 1,000,000. Text accompanying *supra* note 100.

I then multiplied the number of immigrants in each cohort by:

0.973 (1.0000 – 0.027 = 0.973) in each of the initial five years after the immigrants moved to the United States;

0.9825 (1.0000 – 0.0175 = 0.9825) in each of the next (second) five years after the immigrants moved to the United States;

0.9876 (1.0000 – 0.0124 = 0.9876) in each of the third five years after the immigrants moved to the United States;

0.9905 (1.0000 – 0.0095 = 0.9905) in each of the fourth five years after the immigrants moved to the United States.

For a given cohort, the number of lawful immigrants who remained in the United States was computed for the first post-immigration year by multiplying (i) the initial number of lawful immigrants in the cohort and (ii) 0.973. The result of this calculation was multiplied by 0.973 for the second post-immigration year; the calculations continued through the fifth post-immigration year using the result from each preceding calculation multiplied by 0.973. For the sixth through tenth post-immigration years, the result from the immediately prior calculation was multiplied by 0.9825; for the eleventh through fifteenth post-immigration years, the result from the immediately prior calculation was multiplied by 0.9876; and for the sixteenth through the twentieth post-immigration years, the result from the immediately prior calculation was multiplied by 0.9905. The calculations ended with 2010. Accordingly, only one cohort (the cohort of 1990) was followed for twenty years; subsequent cohorts were followed for fewer than twenty years.

The preceding calculation process yielded estimates of the total number of immigrants who lawfully became residents of the United States in each of twenty years — 1990 through 2009 — and who were still in the United States in 2010. Because I assumed that female immigrants had no births while in the United States, my calculation understates the impact of international migration on the size of the U.S. population to the degree that births to immigrant women while in the United States were not numerically offset by deaths of U.S.-present immigrants.

[105] Calculated from the total resident population reported by the decennial census conducted in 1990 and the decennial census conducted in 2010. U.S. CENSUS BUREAU, UNITED STATES SUMMARY:

2010. POPULATION AND HOUSING UNIT COUNTS 1 tbl. 1 (Report CPH-2-1) (2012), *available at* https://www.census.gov/library/publications/2012/dec/cph-2.html.

Chapter 2
An Empirical Study of Americans' Attitudes

I next investigate a possible reason that environmentalists (or at least U.S. environmentalists) currently lack interest in the increasing size of the human population and its impact on nature. Specifically, I explore whether, at the present time, Americans generally exhibit little or no sensitivity to the effect that population growth has on the ecosystem of the planet. If this is the case, American environmentalists will simply be manifesting the attitudes of the public as a whole. A study that used data gathered in 1969 from a nationwide sample of U.S. adults found that attitudes toward the environment were no more than modestly related to attitudes toward population increase.[1] Specifically, in the sample as a whole, respondents who were "deeply" or "somewhat" concerned about environmental quality were more likely than respondents who were "not very concerned" to believe that population size will need to be limited "if our present living standards are to be maintained." However, the relationship between the two attitudes, although statistically significant, was not strong.

Are the attitudes still linked and, if so, to what degree? Chapter 2 is devoted to this question. Importantly, since the earlier study, the statistical tools for analyzing quantitative social science data have made impressive strides, and more-rigorous analyses of data are now possible. Since recent national sample surveys supply data that can be used to ascertain whether the relationship exists among U.S. residents at the present time, Chapter 2 reports another study of the degree to which concern with the environment is tied to concern with mounting population numbers . Because environmental law scholarship needs to be but has rarely been grounded on empirical

[1] Larry D. Barnett, *Concern with Environmental Deterioration and Attitudes toward Population Limitation*, 20 BIOSCIENCE 999, 1000 tbl. 1 (1970). Cramer's V, the measure of association, was 0.131 among all respondents. For an explanation of Cramer's V, see Alan C. Acock & Gordon R. Stavig, *A Measure of Association for Nonparametric Statistics*, 57 SOC. FORCES 1381 (1979).

L. D. Barnett, *Demography and the Anthropocene*,
SpringerBriefs in Environment, Security, Development and Peace 35,
https://doi.org/10.1007/978-3-030-69428-9_2

evidence,[2] the current study may help to refocus environmental law scholarship on population pressure and what might be done to reduce it.

The study that I report here relies on data from the General Social Survey ("GSS"), which interviews probability-selected cross-sections of adults (persons 18 years of age and older) in the United States.[3] The Survey questionnaire in 2000 and in 2010 included two questions (see *infra* Part 2.1 and Part 2.2) that are directly relevant to the focus of the current study. Each question asked interviewees to respond to a statement in terms of whether they "Strongly Agree," "Agree," "Neither Agree nor Disagree," "Disagree," "Strongly Disagree," or "Can't Choose."[4] In the data analysis that is presented below, responses to the statements were measured by dummy variables, i.e., responses were dichotomized and coded as either 0 or 1.[5] The data combine respondents in the 2000 GSS and respondents in the 2010 GSS.

2.1 Dependent Variable

The dependent variable in my study was whether population growth was regarded as a problem. As the measure of the dependent variable, I used responses to the following statement (mnemonic label POPPROB)[6]:

> The earth simply cannot continue to support population growth at its present rate.[7]

---

[2]Robert L. Fischman & Lydia Barbarsh-Riley, *Empirical Environmental Scholarship*, 44 ECOLOGY L.Q. 767, 768, 806 (2018).

[3]NAT'L OPINION RES. CTR., GENERAL SOCIAL SURVEYS, 1972- 2018: CUMULATIVE CODEBOOK viii, 3171–72 (Dec. 2019) [hereinafter CUMULATIVE CODEBOOK], https://gss.norc.org/get-documentation. The GSS was first conducted in 1972. *Id.* at viii. Prior to 2006, GSS samples were limited to persons who spoke English; in 2006 and later years, GSS samples included persons who spoke English and persons who spoke only Spanish. *Id.* at 3171. An overview of the General Social Survey is in TOM W. SMITH, THE GENERAL SOCIAL SURVEYS (GSS Project Report No. 32, 2016), https://gss.norc.org/get-documentation/project-reports.

[4]GSS questionnaires are available at https://gss.norc.org/get-documentation/questionnaires. For the two statements in the 2000 GSS questionnaire, see Nat'l Opinion Res. Ctr., Self-Administered Questionnaire: General Social Survey ([2000]), at 4, 6 [hereinafter 2000 Questionnaire], https://gss.norc.org/get-documentation/questionnaires (under "2000 Questionnaires," select hyperlink "2000 GSS SAQ ISSP"). For the two statements in the 2010 GSS questionnaire, see Nat'l Opinion Res. Ctr., English Questionnaires: Cross-Section 2010 Version 1, at 144, 146 [hereinafter 2010 Questionnaire], https://gss.norc.org/get-documentation/questionnaires (open dropdown menu "2010 Questionnaires;" under "English Questionnaires – Cross-Section," select hyperlink "2010 Version 1").

[5]A "dummy variable" is explained in MELISSA A. HARDY, REGRESSION WITH DUMMY VARIABLES 7–8 (1993).

[6]The mnemonic label that I assigned a variable is not necessarily the mnemonic label assigned by the GSS.

[7]2000 Questionnaire, *supra* note 4, at 4; 2010 Questionnaire, *supra* note 4, at 144. Although for measuring attitudes single-statement instruments are generally less reliable and valid than multi-statement instruments, single-statement instruments are sufficiently reliable and valid to yield acceptable estimates. Marko Sarstedt & Petra Wilczynski, *More for Less? A Comparison of Single-Item and Multi-Item Measures*, 69 DBW 211, 218–19, 223–24 (2009).

In my analysis of the data, I sought to identify respondents who unequivocally expressed concern with population growth, and to contrast them with respondents who did not. Accordingly, interviewees were coded 1 if they responded "Strongly Agree" or "Agree" to the POPPROB statement, and were coded 0 if they responded "Strongly Disagree," "Disagree," "Neither Agree nor Disagree," or "Can't Choose."[8] Because the dependent variable was coded with the digits 0 and 1, i.e., was measured as a dummy, regression models were tested with maximum-likelihood logistic regression.[9]

2.2 Explanatory Variable

The independent variable of chief interest in my study was whether interviewees viewed safeguarding the environment as a priority, and I will focus on the ability of this variable (the "explanatory variable") to account for differences between respondents on the dependent variable. To measure the explanatory variable, I employed responses to the following statement (mnemonic label ENVIMP):

For a five-statement instrument gauging attitudes toward population growth, see Larry D. Barnett, *Women's Attitudes toward Family Life and U.S. Population Growth*, 12 PAC. SOCIOL. REV. 95 (1969). This five-statement instrument has been criticized, but the criticism stemmed from a subjective assessment and facial examination of the instrument. Richard R. Clayton, *Guttman Scaling: An Error Paradigm*, 16 PAC. SOCIOL. REV. 5, 8, 10 (1973). Importantly, the critique did not include any original quantitative data suggesting that the instrument was faulty, and it ignored the quantitative evidence that had been presented backing the conclusion that the five statements in the instrument tap the same attitude. The evidence, which came from two different samples, is in: Barnett, *supra*, at 96 & n.3; Larry D. Barnett, *U.S. Population Growth as an Abstractly-Perceived Problem*, 7 DEMOGRAPHY 53, 56 (1970). Nonetheless, revision of one or more of the five statements in the instrument may be warranted today in order to take account of the change in social context that has occurred during the last half-century. *See generally* EARL BABBIE, SURVEY RESEARCH METHODS 167–70 (2d ed. 1990), which explains the concept (Guttman Scaling) that underlies the evidence supporting the measurement integrity of the five-statement instrument.

[8] On the POPPROB statement, no interviewees were recorded as having selected the "Can't Choose" alternative, but some interviewees were recorded as "Don't Know." I assumed that interviewees who selected the "Can't Choose" alternative were recorded as "Don't Know."

[9] FRED C. PAMPEL, LOGISTIC REGRESSION (2000); Francis L. Huang & Tonya R. Moon, *What Are the Odds of That? A Primer on Understanding Logistic Regression*, 57 GIFTED CHILD Q. 197 (2013).

To analyze the data, I used Stata™ IC version 12.1. The LOGISTIC command was employed for the estimation of odds ratios, and the LOGIT command was used for the estimation of regression coefficients. STATACORP, STATA BASE REFERENCE MANUAL: RELEASE 12, at 932, 970 (2011).

Stata automatically tests the degree of collinearity of every independent variable in a regression model and removes any independent variable whose collinearity is excessive. STATACORP, *supra*, at 975. When testing the models in Table 2.1, Stata did not detect inordinate collinearity and thus kept all of the independent variables. To buttress the decision by Stata not to discard an independent variable, I used the Stata command VIF, UNCENTERED to estimate the Variance Inflation Factor (VIF) score of each independent variable in every model; the highest VIF score (9.0) was below the score (10.0) at which undue collinearity may exist. Further information on collinearity and the Variance Inflation Score is in LARRY D. BARNETT, SOCIETAL AGENTS IN LAW: QUANTITATIVE RESEARCH 15 (2019).

There are more important things to do in life than protect the environment.

In the data analysis, I separated respondents who unambiguously articulated a concern with the environment from respondents who did not. Accordingly, interviewees were coded 1 on the explanatory variable if their response to the ENVIMP statement was "Strongly Disagree" or "Disagree," and were coded 0 on the explanatory variable if their response was "Strongly Agree," "Agree," "Neither Agree nor Disagree," or "Can't Choose."[10] This coding allowed the interviewees for whom the environment was clearly a concern to be contrasted with respondents for whom the environment was not plainly a concern.

2.3 Control Variables

Did differences between respondents on the explanatory variable (ENVIMP) predict differences between respondents on the dependent variable (POPPROB)? In particular, did American adults who were concerned with the environment have, relative to American adults who were not concerned, a greater propensity to be troubled by population growth? To answer the question, the possibility must be taken into account that other variables created, altered, or suppressed a relationship between the explanatory variable and the dependent variable. I investigated the foregoing possibility by holding constant, i.e., controlling, variables that sociological theory suggested could affect the relationship between ENVIMP and POPPROB. The potentially contaminating variables—the control variables in the data analysis—were the culture of the geographic region in which respondents were interviewed; the age, race, and sex of respondents; the educational attainment of respondents; and the size of the population in the area where respondents were interviewed.[11] The location of an interview is presumed to be the residence of the interviewee, because in each sampled household, the GSS confined interviews to one adult among all persons who "usually live" in the household.[12]

A list of the control variables is, of course, incomplete by itself. In particular, such a list needs to be supplemented by (i) an explanation of the theory that supplied the basis for choosing each control variable and (ii) a description of the manner in which each control variable was quantified. I cover the former in Part 2.3.1 and the latter

[10]On the ENVIMP statement, no interviewees were recorded as having selected the "Can't Choose" alternative, but some interviewees were recorded as "Don't Know." I assumed that interviewees who selected the "Can't Choose" alternative were classified as "Don't Know."

[11]Income was omitted as a control variable. The GSS in 2000 and in 2010 included two questions on yearly income (INCOME and RINCOME) that employed the same income categories, but the highest category was "$25,000 or over." CUMULATIVE CODEBOOK, *supra* note 3, at 216–17. Without categories for yearly incomes above $25,000, regression models using GSS data on income were thought to have an unacceptable likelihood of yielding incorrect conclusions. Educational attainment was included as a control variable, however, and is strongly correlated with income. JENNIFER CHEESEMAN DAY & ERIC C. NEWBURGER, U.S. CENSUS BUREAU, THE BIG PAYOFF: EDUCATIONAL ATTAINMENT AND SYNTHETIC ESTIMATES OF WORK-LIFE EARNINGS (2002).

[12]CUMULATIVE CODEBOOK, *supra* note 3, at 2976, 3244–45, 3420–21.

in Part 2.3.2. Readers who want to avoid technical aspects of the study may wish to skip Part 2.3.2 as well as the discussion of interaction variables in Part 2.4. Part 2.3.2 and Part 2.4 can be consulted as needed when considering the data analyses (Part 2.5 and Part 2.7).

2.3.1 Control Variables: Theory

Culture consists of the basic, often-subconscious suppositions that human beings apply to their physical and social surroundings. However, the substance of culture is not uniform across the United States; instead, it differs to some extent from one geographic region of the country to another.[13] Since culture influences cognition,[14] differences in culture have the ability to produce dissimilar thinking about topics such as the environment and population growth.

Cultures can be classified by, inter alia, the degree to which they support unconventionality.[15] All else being equal, a culture that favors unconventionality is likely to be more receptive to ideas not currently accepted than a culture that favors conventionality. On logical grounds, receptivity to ideas that have not previously been considered, or that were considered but rejected, potentially includes a willingness to recognize continued population growth as a problem.

Age, race, and sex. Social divisions in a society are associated with different life experiences, and in the United States, social splits exist along numerous lines, including age, race, and sex (gender). The splits by age, race, and sex have important consequences in American society—e.g., in terms of differences in literacy,[16] income,[17] and social values[18]—that may lead to dissimilarities in interpretations and views.

---

[13] LARRY D. BARNETT, SOCIETAL AGENTS IN LAW: A MACROSOCIOLOGICAL APPROACH 91–98 (2019) [hereinafter SAIL VOL. 1].

[14] Paul DiMaggio, *Culture and Cognition*, 23 ANN. REV. SOCIOL. 253, 282 (1997).

[15] SAIL VOL. 1, *supra* note 13, at 98.

[16] IRWIN S. KIRSCH ET AL., NAT'L CTR. FOR EDUC. STAT., ADULT LITERACY IN AMERICA xviii–xix, 3–6, 8–11, 30–34, 38–39, 46–47, 60–61 (3d ed. 2002), *available at* nces.ed.gov/pubs93/93275.pdf. *See* Roberto M. De Anda & Pedro M. Hernandez, *Literacy Skills and Earnings: Race and Gender Differences*, 34 REV. BLACK POL. ECON. 231, 235, 240 (2007) (finding that the effect of literacy-skill level on the amount of earned income differs by race and by sex).

[17] Emily Greenman & Yi Xie, *Double Jeopardy – The Interaction of Gender and Race on Earnings in the United States*, 86 SOC. FORCES 1217, 1226, 1236 (2008); Andrea E. Willson, *Race and Women's Income Trajectories: Employment, Marriage, and Income Security Over the Life Course*, 50 SOC. PROBS. 87, 91–92, 104 (2003).

[18] Taciano L. Milfont et al., *Values Stability and Change in Adulthood: A 3-Year Longitudinal Study of Rank-Order Stability and Mean-Level Differences*, 42 PERSONALITY & SOC. PSYCHOL. BULL. 572, 575, 586 (2016); Seth Ovadia, *Race, Class, and Gender Differences in High School Seniors' Values: Applying Intersection Theory in Empirical Analysis*, 82 SOC. SCI. Q. 340, 345, 349 tbl. 2, 350 tbl. 3, 353 tbl. 4 (2001).

The *educational attainment* of adults in a society is an empirical indicator of the amount of knowledge that the society is using: An increase in educational attainment signals that more knowledge is being employed. The amount of utilized knowledge in turn affects the degree to which reactions to situations and events are rational.[19] Differences in the extent of rationality are potentially linked to dissimilar perspectives.

The *population size* of the areas where the interviews took place (and the respondents were assumed to reside) was entered as a control, because population size has an obvious potential to affect the degree to which respondents are sensitive to number of people and to increases in that number. Moreover, population size, according to urbanism theory in sociology, has a direct (though not necessarily linear) relationship to the degree of unconventionality in social life: As the number of people in a geographic area becomes larger, unconventionality is expected to be more common.[20]

2.3.2 Control Variables: Measurement

To measure the degree to which American culture endorses unconventionality, I relied on a single dummy variable that was based on the four major geographic regions of the United States that are used by the U.S. Census Bureau, viz., Northeast, South, Midwest, and West.[21] Respondents who resided in the Northeast and respondents who resided in the West were coded 1; the culture in both of these regions places a relatively strong emphasis on unconventionality.[22] Respondents who resided in the South and respondents who resided in the Midwest, on the other hand, were coded 0. Culture in these regions (the South and the Midwest) tends to accentuate conventionality, not unconventionality.[23]

In terms of the bases for social divisions in the United States, age (mnemonic label AGE) was measured by the exact age of the respondent.[24] Sex (mnemonic label SEX) was measured as a dummy variable, with females coded 1 and males coded 0. Race was measured using three categories, viz., "White," "Black," and "Other,"[25] with

---

[19]SAIL VOL. 1, *supra* note 13, at 105.

[20]Claude S. Fischer, *Toward a Subcultural Theory of Urbanism*, 80 AM. J. SOCIOL. 1319, 1324–30 (1975); Claude S. *Fischer, The Subcultural Theory of Urbanism: A Twentieth-Year Assessment*, 101 AM. J. SOCIOL. 543, 544–46 (1995). See, in addition, the text that accompanies *supra* note 15.

[21]In the GSS, the variable REGION reports the geographic division in which each respondent resided. CUMULATIVE CODEBOOK, *supra* note 3, at 233. To create the dummy variable for culture, I combined the geographic divisions in each of the four regions. For the states that form each region, see U.S. BUREAU OF THE CENSUS, GEOGRAPHIC AREAS REFERENCE MANUAL. CH. 6: STATISTICAL GROUPINGS OF STATES AND COUNTIES, at 2 fig. 6-1, 24 tbl. 6-4 (1994), *available at* https://www.census.gov/geo/reference/garm.html (follow "Chapter 6" hyperlink).

[22]SAIL VOL. 1, *supra* note 13, at 98.

[23]*Id.*

[24]Respondents who were 89 years of age or older were coded as 89. CUMULATIVE CODEBOOK, *supra* note 3, at 3241, 3284.

[25]See *id.* at 198 for the method used in GSS interviews to ascertain the race of respondents. *See generally* Aliya Saperstein & Andrew M. Penner, *Racial Fluidity and Inequality in the United States*,

each race category giving rise to a dummy variable. Three dummy variables for race were accordingly created, viz., WHITE, BLACK, and OTHER. On WHITE, respondents who the GSS listed as White were coded 1, and respondents who the GSS listed as Black or Other were coded 0. On BLACK, respondents who the GSS listed as Black were coded 1, and respondents who the GSS listed as White or Other were coded 0. On OTHER, respondents who the GSS listed as Other race were coded 1, and respondents who the GSS listed as White or Black were coded 0. BLACK and OTHER were used in the regression models, making WHITE the reference group.[26]

Educational attainment was based on the number of years of formal schooling completed by the respondents and was quantified by three dummy variables, viz., LOEDUC, MEDEDUC, and HIEDUC:

Variable	coded 1	coded 0
LOEDUC	12 years or less	13 years or more
MEDEDUC	13, 14, or 15 years	less than 13 years or more than 15 years
HIEDUC	16 years or more	15 years or less

LOEDUC thus differentiates respondents who did not go beyond high school from respondents who finished at least one year of college-level coursework.[27] MEDEDUC divides respondents into (i) those who had one, two, or three years of college and separates them from (ii) those who did not complete at least one year of college-level schooling and those who finished a sufficient number of years of college-level coursework to earn a baccalaureate degree or a post-baccalaureate degree. HIEDUC differentiates respondents who presumably received at least a baccalaureate degree from respondents who did not. MEDEDUC was omitted when analyzing the data and hence served as the reference category.

---

118 AM. J. SOCIOL. 676, 687–88, 696–98 & tbls. 2 & 3, 707–08 (2012) (reporting race-specific rates of consistency and inconsistency in the race categorizations of individuals in a large sample of U.S. residents who were studied over a period of twenty-three years).

[26] OTHER was not the reference category for two reasons—(i) being a heterogeneous group, respondents of Other races provide an ambiguous referent; and (ii) being a numerically small group, respondents of Other races may produce unreliable regression coefficients and odds ratios for the relationship of OTHER to the dependent variable. HARDY, *supra* note 5, at 10. As to (ii), among interviewees of Other races, a total of just 77 respondents in the 2000 GSS were coded 0 or 1 on both ENVIMP and POPPROB, and hence were in the data analysis; the corresponding total in the 2010 GSS was 128 respondents. Particularly in 2000, the regression results for OTHER could have been markedly affected by a small change in the total number of respondents of Other races and/or by a small change in the distribution of respondents of Other races on ENVIMP or on POPPROB.

In the two GSS waves, a substantially larger number of Blacks than of Other races were coded 0 or 1 on both ENVIMP and POPPROB. The total number of Blacks so coded was 179 in the 2000 GSS and 228 in the 2010 GSS.

[27] At the college level, the number of years of schooling was based on courses taken for credit at other than a "business college, technical or vocational school." CUMULATIVE CODEBOOK, *supra* note 3, at 3242.

Population size was measured by three dummy variables that were labeled LOPOP, MIDPOP, and HIPOP. The dummy variables were coded using the following ranges of population size:[28]

Variable	coded 1	coded 0
LOPOP	499,999 or less	500,000 or more
MIDPOP	500,000–999,999	less than 500,000 or more than 999,999
HIPOP	1,000,000 or more	less than 1,000,000

The data analysis included LOPOP and HIPOP but not MIDPOP. The reference group, therefore, was composed of respondents in areas that had a population of 500,000–999,999.

2.4 Interaction Variables

Lastly, the possibility exists that the explanatory variable and a control variable did not operate separately. Instead, the explanatory variable (ENVIMP) may have acted in conjunction with a control variable to affect the dependent variable (POPPROB), and the impact of the explanatory variable could have differed between categories or levels of the control variable. We must thus explore whether, in the terminology of statistics, interaction took place between the explanatory variable and a control variable. However, when looking into potential interaction, independent variables are not chosen haphazardly; they are suggested by theory that provides a basis for believing that the selected variables interact.[29] In the instant study, the sex and the race of respondents are promising candidates for interaction with the explanatory variable, because in the United States, sex as well as race create divergent life experiences and social values.[30] The suppression of stereotypes grounded on sex and stereotypes grounded on race are goals of U.S. law for a reason—these stereotypes, being prevalent, have contributed to sex and race differences in perception and interpretation.[31] As a result, whether and how ENVIMP and POPPROB are related may differ between men and women, and they may also differ between racial groups (Whites, Blacks, and Other). To investigate the possibility of such statistical interaction, the numerical values of ENVIMP were multiplied by the numerical values of SEX; they

---

[28]The data are from the GSS variable SIZE. CUMULATIVE CODEBOOK, *supra* note 3, at 235.

[29]T. P. Hutchinson, *Beyond Interaction: Theories*, 40 QUALITY & QUANTITY 869, 870 (2006); JAMES JACCARD, INTERACTION EFFECTS IN LOGISTIC REGRESSION 12–13 (2001); F. David Schoorman et al., *The Role of Theory in Testing Hypothesized Interactions: An Example from the Research on Escalation of Commitment*, 21 J. APPLIED SOC. PSYCHOL. 1338, 1338, 1349, 1353 (1991).

[30]Text accompanying *supra* notes 16–18.

[31]*See* Marcia L. McCormick, *Stereotypes as Channels and the Social Model of Discrimination*, 36 ST. LOUIS U. PUB. L. REV. 19 (2017) (reviewing empirical research on the nature, operation, and consequences of stereotypes, especially sex- and race-based stereotypes).

Table 2.1 Odds ratios from the regression of POPPROB on the independent variables*

Independent variable	Model I: all respondents	Model II: White women	Model III: White men	Model IV: Black women	Model V: Black men
ENVIMP	1.605[c]	1.595[c]	1.831[c]	1.222	2.516[b]
CULTURE	1.159	1.266	1.006	1.386	0.894
AGE	1.004	1.005	1.001	1.010	0.987
SEX	0.769[c]	–	–	–	–
BLACK	0.489[c]	–	–	–	–
WHITE (reference)	–	–	–	–	–
OTHER	0.995	–	–	–	–
LOEDUC	1.179	1.452[b]	0.836	1.144	1.941
MEDEDUC (reference)	–	–	–	–	–
HIEDUC	1.407[c]	2.172[c]	0.871	1.001	0.916
LOPOP	0.683	0.568	0.680	0.885	0.171[a]
MIDPOP (reference)	–	–	–	–	–
HIPOP	0.610[a]	0.429	0.821	1.390	0.103[a]
Intercept	1.004	0.710	1.494	0.239[a]	2.792
Number of respondents	2,587	1,101	876	268	134
Accuracy of predictions (%)	58.2	59.3	57.4	68.3	61.9

* Level of statistical significance of an odds ratio is signified by: [a]≤ .10, [b]≤ .05, [c]≤ .01.
Omission of a variable from a model is signified b y – .

were, multiplied, too, by the numerical values of WHITE, ofBLACK, and ofOTHER.[32] The result was four interaction-testing variables, which were given the mnemonic names ENVSEX, ENVWHITE, ENVBLACK, and ENVOTHER, respectively. ENVWHITE was omitted in the regression analysis because WHITE was omitted for race.[33] Whites, accordingly, were the reference whenever POPPROB was regressed on a set of independent variables that included race or a set of independent variables that included both race and the race-based interaction variables.[34]

2.5 Data Analysis: All Respondents

Several sets of independent variables were utilized in my attempt to account for differences between interviewees in their responses to POPPROB. Each set is by convention designated a "model." The odds ratios that were estimated for the independent variables in five models are shown in Table 2.1. Levels of two-tailed statistical signifi-

[32]JACCARD, *supra* note 29, at 14.

[33]Text accompanying *supra* note 26.

[34]JACCARD, *supra* note 29, at 18–20.

cance were ascertained by first computing an adjusted standard error for each regression coefficient and then dividing the regression coefficient by its adjusted standard error.[35] The adjusted standard errors, which were necessitated by the sampling procedure that the GSS employs,[36] were the mathematical product of 1.19 and the Stata-estimated standard error for each coefficient.[37] The division of a given coefficient by its adjusted standard error determined the z-score, which fixed the significance level, of the coefficient for an independent variable; the significance level of the coefficient was, in turn, the significance level of the odds ratio for the independent variable.

Before we look at Table 2.1, an initial question must be addressed. The question involves the ENVIMP- POPPROB relationship, which is the chief focus of the instant study, and it arises because the data for the study are from two samples that were drawn ten years apart (2000 and 2010). Specifically, the question is whether the magnitude and/or direction of the ENVIMP- POPPROB relationship differed between the two years. To obtain the answer, I constructed a pair of dummy variables—one (mnemonic label YEAR) for the year in which the survey was done, and the other (mnemonic label ENVYEAR) for possible interaction between ENVIMP and YEAR.[38] I then regressed POPPROB on ENVIMP, CULTURE, AGE, SEX, BLACK, OTHER, LOEDUC, HIEDUC, LOPOP, HIPOP, YEAR, and ENVYEAR. I n this regression, the odds ratio for the

---

[35]In the instant study, the standard error is the standard deviation of the "sampling distribution" for the regression coefficient, i.e., the standard deviation of the dispersion that would have been obtained for the regression coefficient across numerous probability samples of adults who resided in the United States (the universe being studied). The sampling distribution assumes that, in the universe, the regression coefficient for a given independent variable is zero and that the independent variable is thus unrelated to the dependent variable. This assumption is commonly known as the null hypothesis.

The sampling distribution supplies the basis for ascertaining the probability that the numerical value of a particular regression coefficient is due solely to chance when the null hypothesis is correct, i.e., when the regression coefficient is zero in the universe. The probability, which is the level of statistical significance for the regression coefficient, represents the likelihood of making a mistake by rejecting the null hypothesis. The highest generally accepted probability is 10% (the .10 significance level).

[36]GSS interviewees reside in housing units that are selected using multi-stage cluster sampling. CUMULATIVE CODEBOOK, *supra* note 3, at 3171–72. Statistics software, when applied to data obtained through cluster sampling, produces an artificially low estimate of the standard error for a regression coefficient. Lin Wang & Xitao Fan, The Effect of Cluster Sampling Design in Survey Research on the Standard Error Statistic (paper presented at the 1997 meeting of the American Educational Research Ass'n), *available at* https://eric.ed.gov/?id=ED409320. However, while the need to correct standard errors arising from cluster sampling is widely acknowledged, every procedure for making the correction (including the procedure employed here) has disadvantages. Andrew Gelman, *Struggles with Survey Weighting and Regression Modeling*, 22 STAT. SCI. 153, 163–64 (2007).

For a brief explanation of cluster sampling as a form of probability sampling, see BABBIE, *supra* note 7, at 87–91.

[37]Larry D. Barnett, *Mutual Funds, Hedge Funds, and the Public-Private Dichotomy in a Macrosociological Framework for Law* 33-34 (CIRSDIG Working Paper No. 34, 2008), https://papers.ssrn.com/sol3/papers.cfm?abstract_id=1319782.

[38]The interaction variable was created by multiplying, for each respondent, the numerical code assigned for ENVIMP and the numerical code assigned for YEAR. JACCARD, *supra* note 29, at 14.

interaction variable (ENVYEAR) was not statistically significant at a probability of 0.10 or below. I therefore did not reject the null hypothesis, i.e., I concluded that, in 2000 and in 2010, the relationship between ENVIMP and POPPROB was the same, in direction and degree, among U.S. adults as a whole.[39] Survey year, in short, could be disregarded in the remaining regressions, and was.

Turning to Table 2.1, the initial model—Model I—reveals that ENVIMP, SEX, BLACK, HIEDUC, and HIPOP had odds ratios that were statistically significant at or below a probability of 0.10.[40] A helpful approach to interpreting the odds ratio for an independent variable is to subtract 1.000 from the odds ratio and multiply the result by 100.[41] In the instant study, the calculation yields the percentage by which the odds of viewing population growth as a problem undergoes change for an individual who moves up one measurement category or one measurement unit on the independent variable under consideration while staying the same on the other independent variables in the model.[42] Thus, as to the explanatory variable (ENVIMP), the odds that population growth would be considered a problem were 60.5% higher when, with the other independent variables held constant, an individual believed that protecting the environment *was* important than when the individual felt that protecting the environment was *not* important. The odds ratios for two of the control variables (SEX and BLACK) should also be noted: All else being equal, the odds of viewing population growth as problematic were 23.1% lower for a woman than for a man and 51.1% lower for a Black than for a White.

A final matter regarding Model I is the degree to which it fit the data. As seen in the bottom row of the table, the fit of Model I was modest: Model I was able to predict accurately the category of the dependent variable (POPPROB) in which 58.2%, or roughly six out of ten, of the respondents fell.[43] Given that the dependent variable was measured using two categories, accurate predictions would have been made for half of the respondents if the distribution of respondents across these categories was

---

[39]Maarten L. Buis, *Stata Tip 87: Interpretation of Interactions in Nonlinear Models*, 10 STATA J. 305 (2010). The regression coefficient for the interaction variable (ENVYEAR) had a z-score (calculated with its adjusted standard error) of 0.70; the significance level of the regression coefficient, therefore, was 0.49. Since every respondent in this model had a Cook's Statistic below 0.11, none of the respondents appeared to be an influential outlier. DAVID W. HOSMER & STANLEY LEMESHOW, APPLIED LOGISTIC REGRESSION 173, 180 (2d ed. 2000); J. SCOTT LONG & JEREMY FREESE, REGRESSION MODELS FOR CATEGORICAL DEPENDENT VARIABLES USING STATA 151 (2d ed. 2006).

[40]In Model I, Cook's Statistic for every respondent was negligible (less than 0.10), indicating that no influential outlier was present.

[41]*See* Long & Freese, *supra* note 39, at 178, 180 (discussing odds ratios (signified by e^b) and percentage change in odds (signified by %)).

[42]The odds that an individual has a particular attribute is the ratio of (i) the probability that the attribute is present (i.e., the proportion of individuals who possess the attribute) to (ii) the probability that the attribute is absent (the proportion of individuals who do not possess the attribute). The attribute here is the belief that population growth is a problem.

[43]The accuracy rate of predictions as a measure of model fit is discussed in Long & Freese, *supra* note 39, at 104, 110–12. The rate is obtained with the ESTAT CLASSIFICATION command in Stata. STATACORP, *supra* note 9, at 957.

unknown and respondents had been randomly placed in the categories. Under the foregoing assumption, the independent variables in Model I improved the accuracy of predictions by roughly eight percentage points, or about 16%.

Model I was then probed for possible statistical interaction between its explanatory variable and either sex or race. The probe, whose results are not reported in Table 2.1 but are summarized here, was conducted by regressing POPPROB on each of the independent variables in Model I *and* the interaction variables ENVSEX, ENVBLACK, and ENVOTHER.[44] Collectively, the expanded group of variables (i.e., the Model I independent variables joined by the three interaction variables) was not more accurate than just the independent variables of Model I in placing respondents in the categories of the dependent variable.[45] The degree to which the expanded group of independent variables fit the data, in other words, was no better than the degree to which Model I fit the data. Additionally, none of the three interaction variables had an odds ratio that was statistically significant at or below a probability of 0.10. There was, accordingly, no evidence that the explanatory variable interacted with sex or with race.[46]

2.6 Intersectionality Theory

Despite the statistical non-significance of the odds ratios for the three interaction variables, our consideration of sex and race is not at an end. The interaction variables revealed that the ENVIMP- POPPROB relationship did not differ between women and men or between Blacks and Whites, but each interaction variable involved either sex *or* race rather than sex *and* race. Each interaction variable thus utilized only a single trait. However, intersectionality theory, which rejects any approach to understanding social life that is based on "single-axis thinking,"[47] suggests that sex and race should be considered jointly rather than singly. According to intersectionality theory, social life in a structurally complex society is grounded on, inter alia, various individual-level traits that are entwined and that work in concert to form social categories each of which confers a distinct social identity on its members.[48] To illustrate, let us assume

---

[44]The regression with the expanded set of independent variables omitted ENVWHITE because WHITE was the reference for race. See text accompanying *supra* note 26.

[45]The accuracy rate for the expanded group of variables was 57.9%. As seen in the bottom row of the table, the accuracy rate for Model I was 58.2%.

[46]Jaccard, *supra* note 29, at 18–23.

[47]Sumi Cho et al., *Toward a Field of Intersectionality Studies: Theory, Applications, and Praxis*, 38 SIGNS 785, 787 (2013). In addition, see Kimberle Crenshaw, *Demarginalizing the Intersection of Race and Sex: A Black Feminist Critique of Antidiscrimination Doctrine, Feminist Theory and Antiracist Politics*, 1989 U. CHI. LEGAL F. 139.

[48]Cho et al., *supra* note 47, at 787, 797; Mike C. Parent et al., *Approaches to Research on Intersectionality: Perspectives on Gender, LGBT, and Racial/Ethnic Identities*, 68 SEX ROLES 639, 640 (2013).

As originally formulated, intersectionality theory focused on social conflict and included a political agenda. Cho et al., *supra* note 47, at 797, 800. However, intersectionality theory is likely to be more productive as a heuristic tool in empirical social science research when it does not emphasize social conflict and does not have a political agenda than when it does. Notably, a longstanding theory

that social life in a society is organized around two traits (arbitrarily labeled trait #1 and trait #2) and that both traits consist of dichotomous categories/identities. If the categories/identities for trait #1 are designated A and B and the categories/identities for trait #2 are designated C and D, the social categories/identities obtained from the mixture of the two traits are AC, AD, BC, and BD. Intersectionality theory focuses on the categories/identities arising from combinations of trait #1 and trait #2 ("multiple axis categories/identities"), not on the categories/identities of trait #1 and of trait #2 separately ("single axis categories/identities"). Intersectionality theory, in other words, is concerned with AC, AD, BC, and BD rather than with A, B, C, and D individually. For the instant study, the theory directs attention to the categories/identities of White women, White men, Black women, and Black men. The theory looks at the foregoing because it considers each of their component groups—women, men, Whites, and Blacks—to be heterogeneous, not homogeneous.

Intersectionality theory has been elaborated using the concepts of "switch" intersectionality and "nonadditive" intersectionality. Switch intersectionality concentrates on *whether* a multiple-axis category/identity in a set of such categories/identities is affected (i.e., turned on) by a given situation or setting; nonadditive intersectionality concentrates on the *strength* of the effect on the members of each multiple-axis category/identity that is affected.[49] I discuss each of these concepts in turn.

Switch intersectionality posits that the impact of a particular situation or setting is triggered for at least some multiple-axis social categories/identities in a class of such categories/identities, but that the impact may not be triggered for every multiple-axis category/identity in the class. Switch intersectionality thus maintains that the social categories/identities from a fusion of traits may differ among themselves in whether they are touched by a situation or setting. To use the two-trait illustration above, switch intersectionality is exemplified when a situation or setting does not affect the members of one multiple-axis category/identity (say AC) but does affect the members of the other three multiple-axis categories/identities (AD, BC, and BD). In this case, the members of AC are unaffected by the situation or setting while the members of AD, BC, and BD are.

Nonadditive intersectionality supplements switch intersectionality and is germane to an analysis of social life once switch intersectionality is taken into account. Nonadditive intersectionality posits that a situation or setting has a larger impact or a smaller impact on the members of an affected multiple-axis category/identity than it has on the members of each of the component single-axis categories/identities. Where nonadditivity occurs, the size of the effect of a situation/setting on a multiple-axis category/identity is a nonlinear arithmetic function of the effect of the situation/setting on the single-axis categories/identities that comprise the multiple-axis

in sociology—structural functionalism—(i) posits that social equilibrium normally takes priority over social conflict in a society and (ii) omits a political agenda. Part 3.2 in *infra* Chap. 3.

[49]Liam Kofi Bright et al., *Causally Interpreting Intersectionality Theory*, 83 PHIL. SCI. 60, 62 (2016).

category/identity.[50] To continue with the assumption that among the four multiple-axis categories/identities only AC is unaffected, nonadditivity means that, in terms of the magnitude of the impact of a situation or setting on people, $AD \neq A + D$, $BC \neq B + C$, and $BD \neq B + D$.

2.7 Data Analysis: Sex-Race Groups

Intersectionality theory is relevant here because it assumes that sex and race, as single-axis person-level traits, operate jointly rather than separately, and thereby generate two-axis social categories/identities.[51] Given switch intersectionality and nonadditive intersectionality, meaningful dissimilarities in the ENVIMP- POPPROB relationship can be expected between the sex-race categories/identities of White women, White men, Black women, and Black men. Importantly, well-designed quantitative research on the life situations and trajectories of these sex-race groups provides evidence backing intersectionality theory.[52] As a result, a foundation exists for the hypothesis that each of the four groups will not have the same perspective on the physical and social worlds that all of them face. The hypothesis is certainly plausible as applied to the United States since American society is structurally complex and socially fragmented along lines of sex as well as race.

To test the hypothesis, I ran four sex-by-race regressions of POPPROB on the Model I independent variables other than sex and race. In particular, a regression was done for White women (designated Model II in Table 2.1), for White men (Model III), for

---

[50] *Id.* at 63.

[51] Amy C. Steinbugler et al., *Gender, Race, and Affirmative Action: Operationalizing Intersectionality in Survey Research*, 20 GENDER & SOC'Y 805, 807–08 (2006).

[52] Kerryn E. Bell, *Young Adult Offending: Intersectionality of Gender and Race*, 21 CRITICAL CRIMINOLOGY 103 (2013) (studying whether members of the birth cohort born in 1958 in a major U.S. city had crime-related police contacts in that city or in the metropolitan area where that city is located; using, for each of its six dependent variables, the frequency with which members of this cohort had crime-related police contacts as juveniles and as young adults through age 26; finding that White females, Black females, and Black males generally differed from White males on the dependent variables; and concluding that sex and race interact in determining the incidence of crime-related police contacts by juveniles and young adults); Huoying Wu, *Can the Human Capital Approach Explain Life-Cycle Wage Differentials Between Races and Sexes?* 45 ECON. INQUIRY 24, 34 tbl. 5, 35–36 (2007) (using data from a national probability sample of U.S. residents who were 14–21 years old when initially interviewed in 1979 and who were re-interviewed in 1994; studying the interviewees from age 20–21 to age 31–33; finding differences during the course of aging between White men, White women, Black men, and Black women on several economic measures, viz., mean number of weeks employed and unemployed, mean number of weeks out of the labor force, and mean number of hours worked weekly). *Accord*, Lauren B. Edelman et al., *Legal Discrimination: Empirical Sociolegal and Critical Race Perspectives on Antidiscrimination Law*, 12 ANN. REV. L. & SOC. SCI. 395, 401–02 (2016). *See also* Glenn Firebaugh et al., *Why Lifespans Are More Variable Among Blacks Than Among Whites in the United States*, 51 DEMOGRAPHY 2025, 2027, 2032–33, 2037–38 & fig. 7, 2041 (2014) (using data on the resident population of the United States; finding that age at death is more variable among Blacks than among Whites who die of the same cause; and concluding that this variability is mainly due to age at death among females and secondarily due to age at death among males).

Black women (Model IV), and for Black men (Model V).[53] Model II through Model V thus allow us to look within and across sex-race groups.

Perhaps the most salient finding from the four sex-by-race models is that not all of the models had an odds ratio for ENVIMP that was statistically significant. Although the odds ratio for ENVIMP exceeded 1.000 in every model, it was not statistically significant at or below 0.10 in the model for Black women. Consequently, among Black men, White men, and White women—whose odds ratios for ENVIMP reached statistical significance at or below .05—concern with the environment was associated with higher odds that population growth would be seen as a problem: Sensitivity to the environment lifted the odds of regarding population growth as problematic by approximately 152% among Black men, 83% among White men, and 60% among White women.[54] In terms of fitting the data (bottom row of Table 2.1), the models for these groups fell within a narrow range, with the model for Black men having the highest accuracy rate (61.9%) and the model for White men having the lowest (57.4%).

Chapter 2 should not be ended without mentioning a final point. Among the residents of countries other than the United States, we do not know whether there is a relationship between sensitivity to environmental matters and sensitivity to human population increase. Proof of such a relationship outside the United States requires information obtained in national probability-sample surveys from residents of multiple countries, but such data have either not been gathered or, if gathered, have not been analyzed.[55] Future research, accordingly, should explore whether, among residents of non-U.S. countries, concern with the environment predicts concern with population growth. This research may identify aspects of societal structure and culture that create differences in the relationship between attitudes toward the environment and attitudes toward increases in population.

Appendix: Model V

Each of the three Black respondents who were deemed to be influential outliers and hence excluded from the results for Model V had a Cook's Statistic that, although less than 1.00, was relatively large.[56] Notably, all three respondents were elderly and

[53]Regressions were not done for respondents whose race was Other. One reason for this omission is that, in Model 1, the odds ratio for OTHER was not statistically significant. In addition, see *supra* note 26.

[54]The results for Model V (Black men) omit three Black men who were considered to be influential outliers. See *infra* Appendix to this chapter.

[55]Research on the relationship in Western Europe is needed given the ecological footprint of that region (*supra* Figure 1.3 and accompanying text). However, none of the eight rounds of the European Social Survey has obtained information that allows a quantitative test of whether, among residents of participating countries, concern with the environment is related to concern with population growth. The questionnaires for these rounds are available at European Social Survey, Data and Documentation by Round/Year, https://www.europeansocialsurvey.org/data/round-index.html.

[56]Respondents were dropped only when estimating Model V. The data for Model II, Model III, and Model IV did not appear to contain an influential outlier. The maximum Cook's Statistic for a

had little formal schooling,[57] characteristics that could have reduced their ability to understand, and hence could have affected their responses to, the ENVIMP statement (on which they were in the category coded 0) and the POPPROB statement (on which they were in the category coded 1). In addition, the three respondents were in the same GSS sample (viz., 2010), and no other Black male in that sample who was of similar age and educational attainment answered ENVIMP and POPPROB in the way that the three respondents did, i.e., indicated that they were unconcerned with the environment but were concerned with continued population growth.[58] Because the three interviewees were unique in the 2010 sample and had a Cook's Statistic that was twice the size of the Cook's Statistic for any other Black male, they were treated as influential outliers and were excluded from the data used to estimate the odds ratios in Model V.[59]

---

respondent was 0.18 in Model II, 0.18 in Model III, and 0.20 in Model IV. As to Model I, see *supra* note 40.

Each of the three Black male respondents omitted from the data for Model V had a Cook's Statistic of 0.94 while all of the remaining Black male respondents in the data had a Cook's Statistic under 0.47. The three omitted respondents, therefore, met one of the criteria recommended for deciding whether a case should be examined to ascertain if it is an influential outlier — the Cook's Statistic for each of the three omitted respondents was much larger than the Cook's Statistic for other respondents. HOSMER & LEMESHOW, *supra* note 39, at 180; Long & Freese, *supra* note 39, at 151.

[57]The three respondents reported their age as 69 and their years of formal schooling as ten or less. Two of the three respondents reported that they had no more than five years of formal schooling.

[58]The three respondents were located in the South. However, among Black men in all geographic regions who were interviewed in 2010, who in that year were aged 65 or older, and who had less than twelve years of formal education, the three respondents were the only interviewees to be in the category coded 0 on ENVIMP and in the category coded 1 on POPPROB.

[59]With the three respondents in the data, the z-score of the regression coefficient for ENVIMP among Black men ($n = 137$), computed with the adjusted standard error, was 1.77 (regression coefficient = 0.807, odds ratio = 2.242). The significance level of a z-score = 1.77 is .08 and hence $\leq .10$. Without the three respondents in the data, the z-score of the regression coefficient for ENVIMP among Black men ($n = 134$), computed with the adjusted standard error, was 1.98 (regression coefficient = 0.922, odds ratio = 2.516); the significance level of a z-score = 1.98 is $\leq .05$.

Chapter 3
Environmentalism and Interdisciplinarity

Although among Americans sensitivity to the environment is a predictor of sensitivity to human population growth, it is an imperfect predictor, and a commitment to the environment does not automatically produce a recognition that population growth has negative environmental effects. Indeed, in the two GSS samples that provided the data for the study in Chap. 2, only six out of ten White men, White women, and Black men who were concerned with the environment were also concerned with population growth. Given the absence of a close connection between the two concerns, ecosystem deterioration will often not be traced to, and will often not be seen as a function of, the mounting number of human beings. Among Americans, the situation should perhaps not be surprising because environmental concepts and movements in the United States have historically been insular and had a narrow focus.[1] U.S. environmental scholars are likely affected by this context, but they should not be since they are supposed to look at *all* causes of harm to the ecosystem. The present failure of U.S. environmental scholars to address the population factor is, in a word, curious.

One possible reason that U.S. environmental scholars do not confront overpopulation may be that these scholars want to avoid the troubling ethical issues raised by human population growth and its control.[2] Since the ethical issues have no easy solutions, a desire to avoid them is understandable. However, the evidence in Part 1.4 of *supra* Chap. 1 cannot be simply dismissed; indeed, given this evidence, neglect of the population factor is short-sighted. Another possible reason may be that the declining

---

[1]Greg Mitman, In Search of Health: Landscape and Disease in American Environmental History, 10 ENVTL. HIST. 184, 185 (2005).

[2]Dean Spears, *Making People Happy or Making Happy People? Questionnaire-Experimental Studies of Population Ethics and Policy*, 49 SOC. CHOICE & WELFARE 145, 168 (2017). *Accord*, Noah Scovronick et al., *Impact of Population Growth and Population Ethics on Climate Change Mitigation Policy*, 114 PROC. NAT'L ACAD. SCI. 12,338, 12,342 (2017) (contrasting average utilitarianism and total utilitarianism); Veit Bader, *The Ethics of Immigration*, 12 CONSTELLATIONS 331 (2005).

L. D. Barnett, *Demography and the Anthropocene*,
SpringerBriefs in Environment, Security, Development and Peace 35,
https://doi.org/10.1007/978-3-030-69428-9_3

annual rate of world population growth[3] has made environmentalists complacent and led them to believe that overpopulation is no longer a problem. Rates, of course, should not distract from absolute numbers. As Fig. 1.1 (*supra* Chap. 1) demonstrates, a large numerical expansion of the human species has been occurring on the planet for a long time.

3.1 Attitude-Behavior Congruence

Among individuals whose attitudes regarding the environment are related to their attitudes regarding population growth, is the relationship of any practical value? For the question to have an affirmative answer, attitudes on a topic must result in behavior pertinent to the topic. Studies by psychologists that have focused on whether and the extent to which attitudes are manifested in relevant behavior have uncovered a modest, but not strong, overall relationship between attitudes and attitude-pertinent action.[4] To be exact, variation among individuals in attitudes evidently accounts for roughly one-fourth of the variation in attitude-related behavior.[5] Concern with the environment, therefore, is not necessarily, or even usually, converted into action that is intended to slow down or stop increases in population. Nevertheless, concern with the environment supplies a push for action on population growth.[6] Notably, attitude-behavior congruency has been found to increase when the attitudes pertain to an object that becomes more consequential to the individuals who hold the attitudes, to persons in whom these individuals are psychologically invested, or to both.[7] Attitude-behavior congruency has also been found to increase when emotion is attached to

---

[3]Note 28 in *supra* Chap. 1.

[4]In the relationship, attitudes are generally the cause rather than the effect of behavior. Michael Riketta, *The Causal Relation Between Job Attitudes and Performance: A Meta-Analysis of Panel Studies*, 93 J. APPLIED PSYCHOL. 472, 476 (2008).

[5]Laura R. Glasman & Dolores Albarracin, *Forming Attitudes That Predict Future Behavior: A Meta-Analysis of the Attitude-Behavior Relation*, 132 PSYCHOL. BULL. 778, 814 (2006) (reporting a weighted mean correlation coefficient of .52). Since $(0.52)^2 = .27 \times 100 = 27\%$, variance in attitudes explains 27% of the variance in behavior.

[6]In a nation that is structurally complex and democratically governed, macro-level conditions and forces shape the content of law on key social matters. LARRY D. BARNETT, EXPLAINING LAW: MACROSOCIOLOGICAL THEORY AND EMPIRICAL EVIDENCE 8–11, 47–48 (2015) [hereinafter EXPLAINING LAW];LARRY D. BARNETT, SOCIETAL AGENTS IN LAW: A MACROSOCIOLOGICAL APPROACH ch. 2 pt. 2.3 (2019) [hereinafter SAIL VOL. 1]. Macro-level conditions and forces mold the content of law probably by generating, inter alia, attitudes that favor particular law content. *See* Adam J. Zolotor & Megan E. Puzia, *Bans against Corporal Punishment: A Systematic Review of the Laws, Changes in Attitudes and Behaviours*, 19 CHILD ABUSE REV. 229, 242 (2010) (finding that the enactment of legislation prohibiting corporal punishment was preceded by a decline in attitudes favorable to corporal punishment).

[7]Ian M. Johnson et al., *Expanding the Reach of Vested Interest in Predicting Attitude-Consistent Behavior*, 9 SOC. INFLUENCE 20, 21–22, 33 (2014).

the behavior and is involved in prompting it.[8] Efforts such as educational programs that promote personal sensitivity to the environment and to the environmental repercussions of population growth may thus be beneficial in bringing about action on population growth.[9] Such efforts may also create, among individuals who have not recognized the population-environment connection, an appreciation of the damage that population size can inflict on the ecosystem o f he planet.

Assuming that concern with the biosphere leads to government action designed to regulate the size of the human population, let us consider what social science can tell us about whether such action, if it occurs, is likely to be effective. I begin (*infra* Part 3.2) by describing a social science theory that is helpful in understanding the conclusion to be derived (in *infra* Part 3.3) from the findings of pertinent empirical research.

3.2 Structural-Functionalism Theory

Fertility and migration are demographic phenomena, but they occur in a social context, not in a social vacuum. Levels of fertility and migration, accordingly, are determined by societal properties, and an attempt to alter the level of either or both phenomena must take these properties into account. Insight into the properties and how they pertain to fertility and migration is offered by the sociological theory of structural functionalism, which became the reigning theory in American sociology during the 1950s.[10] Although competing sociological theories have attracted adherents—and created serious schisms within sociology[11]—structural functionalism continues to influence sociology and to prove useful in analyzing and explaining social life.[12]

---

[8]Jie Zhou et al., *How Affectively-Based and Cognitively-Based Attitudes Drive Intergroup Behaviours: The Moderating Role of Affective-Cognitive Consistency*, 8(11) PLOS ONE, Nov. 14, 2013, at 1, 10, https://journals.plos.org/plosone.

[9]A study using a nonprobability sample of 300 individuals residing in the United Kingdom found that negative affect regarding population growth, but not exposure to information about population growth, raised the likelihood that respondents saw risk in world population growth; the study also found, however, that negative affect and exposure to information were unrelated to the likelihood of favoring measures to limit population growth. Ian G. J. Dawson & Johnnie E. V. Johnson, *Does Size Matter? A Study of Risk Perceptions of Global Population Growth*, 37 RISK ANALYSIS 65, 67–68, 71, 75 tbl. IV, 77 tbl. V (2017).

[10]John Holmwood, *Functionalism and Its Critics*, in 2 HISTORICAL DEVELOPMENTS AND THEORETICAL APPROACHES IN SOCIOLOGY 110, 110–12 (Charles Crothers ed., 2010), *available at* https://www.eolss.net/Sample-Chapters/C04/E6-99A-26.pdf.

[11]Jonathan H. Turner, *The Disintegration of American Sociology*, 32 SOCIOL. PERSP. 419, 429 (1989). *See also* Reba Rowe Lewis, *Forging New Syntheses: Theories and Theorists*, 22 AM. SOCIOLOGIST 221 (1991) (contending that, in order for sociology to improve the quality of its research and its standing as an academic discipline, the advocates of different theories in sociology must curtail their internecine fights and work toward a unification of their theories).

[12]*E.g.*, Benjamin Cornwell & Edward O. Laumann, *If Parsons Had Pajek: The Relevance of Midcentury Structural-Functionalism to Dynamic Network Analysis*, 17 J. SOC. STRUCTURE 1, 2 (2016),

What are the chief propositions of structural-functionalism theory in terms of the present discussion? A society, according to the theory, is a system[13] that has structural elements as well as functional requirements.[14] The structural elements are societal institutions, i.e., macro-level arrangements such as government and law that are continuing and socially accepted.[15] The functional requirements of a society presumptively include a minimum degree of internal integration,[16] a significant consequence of which is a societal tendency toward inertia.[17] Societies, in the view of structural-functionalism, are thus "bounded systems with certain basic 'societal prerequisites' in the absence of which they would disintegrate."[18] Structural-functionalism theory further contends that interpersonal bonds within a society tie the individual members of a society to one another and that the morphological elements (institutions) of a society largely fit together: Social integration and system integration, in other words, are present in a society.[19] As a result, a society tends to be self-sustaining and self-reinforcing, and the social behaviors central to it are resistant to modification.[20]

---

https://www.cmu.edu/joss/content/articles/volindex.html; Stacie E. Goddard & Daniel H. Nexon, *Paradigm Lost? Reassessing* THEORY OF INTERNATIONAL POLITICS, 11 EUR. J. INT'L REL. 9, 10 (2005); LARRY D. BARNETT, THE PLACE OF LAW: THE ROLE AND LIMITS OF LAW IN SOCIETY 302–310 (2011) [hereinafter THE PLACE OF LAW]. *See also* Ruth Lane, *Structural-Functionalism Reconsidered: A Proposed Research Model*, 26 COMP. POL. 461 (1994) (contending that structural-functionalism and its insights are relevant to and useful in studies of comparative politics). *Cf.* Donald W. Harper, Structural-Functionalism: Grand Theory or Methodology?, at 1 (2011) (arguing that structural-functionalism theory has had, and continues to have, a major influence on scholarly work in the fields of sociology, organizations, and management), https://www.scribd.com/document/261889937/Structural-Functionalism-Harper.

13 THE PLACE OF LAW, *supra* note 12, at 394.

14 Goddard & Nexon, *supra* note 12, at 18.

15 THE PLACE OF LAW, *supra* note 12, at 394.

16 Gösta Carlsson, *Reflections on Functionalism*, 5 ACTA SOCIOLOGICA 201, 201 (1962) (observing that structural-functionalism theory posits the existence of societal needs, including the need for agents that foster cohesiveness in social life).

17 ROBERT K. MERTON, SOCIAL THEORY AND SOCIAL STRUCTURE 176 (1968 enlarged ed.) (contending that strains within a society "exert pressure for change. When social mechanisms for controlling them are operating effectively, these strains are kept within such bounds as to limit change of the social structure.").

18 S. N. Eisenstadt, *Functional Analysis in Anthropology and Sociology: An Interpretative Essay*, 19 ANN. REV. ANTHROPOLOGY 243, 244 (1990).

19 David Lockwood, *Social Integration and System Integration*, *in* EXPLORATIONS IN SOCIAL CHANGE 244, 245 (George K. Zollschan & Walter Hirsch eds., 1964). Within a given society, the degree of social integration and the degree of system integration may vary over time. This fluctuation can result in a difference between the former and the latter at a specific point in time, i.e., at any one time in a society, the degree of social integration may be higher or lower than the degree of system integration. Margaret Archer, *Social Integration and System Integration: Developing the Distinction*, 30 SOCIOL. 679, 693–94 (1996).

20 Nancy Kingsbury & John Scanzoni, *Structural-Functionalism*, in SOURCEBOOK OF FAMILY THEORIES AND METHODS: A CONTEXTUAL APPROACH 195, 196 (P. G. Boss et al. eds., 1993);

Structural-functionalism theory is the foundation for the framework employed here.[21] That framework focuses on social behaviors that are important to the societal system, and at least in the United States, such behaviors have long included marriage and childbearing as an accompaniment of marriage.[22] Although the grip of marriage on the U.S. population weakened during the last half of the twentieth century,[23] marriage still has a strong hold on the majority of Americans: Indeed, three out of four adults in the United States are likely to marry before their fifty-fifth birthday.[24] Moreover, among U.S.-resident women who were born in the twentieth century and who exited their reproductive period prior to the twenty-first century, no less than three-fourths had given birth to a child.[25] While births occurring outside of marriage became a larger share of all births in the United States after 1940 and constituted about two-fifths of total births in the country during the second decade of the twenty-first century, a maximum of around only one in twenty unmarried, childbearing-age women had a non-marital birth in any single year during this time.[26] Equally telling, fewer than three out of ten U.S. women during the twentieth century had, through age 30, borne a child out-of-wedlock despite the long-term increase in the fraction of U.S. women who had experienced a non-marital birth.[27]

LARRY D. BARNETT, SOCIETAL AGENTS IN LAW: QUANTITATIVE RESEARCH 20 (2019) [hereinafter SAIL VOL. 2].

[21] The full framework is presented in EXPLAINING LAW, *supra* note 6, at 7–17.

[22] SAIL VOL. 1, *supra* note 6, at 21, 64 n.126, 85. Immigration and emigration are likely to be society-important activities, too, since they can alter the culture and structure of a society. SAIL VOL. 2, *supra* note 20, at 240–42.

[23] SAIL VOL. 1, *supra* note 6, at 23–26.

[24] KIM PARKER ET AL., PEW RES. CTR., RECORD SHARE OF AMERICANS HAVE NEVER MARRIED 12 (2014) (forecasting the incidence of marriage among U.S. residents born during the period from the mid-1970s to the mid-1980s when, in 2030, this cohort will be in the age range 45–54), https://www.pewsocialtrends.org/2014/09/24/record-share-of-americans-have-never-married.

[25] Donald T. Rowland, *Historical Trends in Childlessness*, 28 J. FAM. ISSUES 1311, 1314–15 tbl. 1, 1323, 1325 Fig. 2 (2007) (using data on the eleven five-year cohorts of females who were born in the twentieth century and reached age 45–49 during the twentieth century, i.e., the 1900–1904 birth cohort through the 1950–1954 birth cohort).

[26] CARMEN SOLOMON- FEARS, CONG. RES. SERV., NONMARITAL BIRTHS: AN OVERVIEW 24–26 tbl. A-1 (2014); Joyce A. Martin et al., Nat'l Ctr. Health Stat., *Births: Final Data for 2018*, NAT'L VITAL STAT. REP., Nov. 27, 2019, at 25 tbl. 9, 26 tbl. 10.

[27] Lawrence L. Wu, *Cohort Estimates of Nonmarital Fertility for U.S. Women*, 45 DEMOGRAPHY 193, 194, 199, 201 tbl. 6 (2008) (studying a set of birth cohorts of U.S. women; finding that, from the first (pre-1925) birth cohort to the last (1965–69) birth cohort that had reached age 30 by June 1995, the percentage of women having had a non-marital birth prior to their thirty-first birthday rose among all women, among White women, and among Black women but that, in the 1965–1969 cohort, the percentage was 26.9 for all women and 19.6 for White women, although it was 61.2% for Black women). In following women through just age 30, Professor Wu notes that the percentage of women bearing a child outside of marriage rises "only modestly" after age 30. *Id.* at 199.

In the United States, the percentage of all births in the 1960–1992 period that were non-marital steadily increased among both Blacks and Whites but was higher among Blacks than Whites. Over the course of 1960–1992, the demographic reasons for the increases in the percentage among Blacks were not identical to the demographic reasons for the increases in the percentage among Whites. Herbert L. Smith et al., *A Decomposition of Trends in the Nonmarital Fertility Ratios of Blacks and*

Marriage and childbearing within marriage, in short, have been significant societal activities in the United States, and they remain so.[28] As to the frequency of such activities, structural-functionalism theory warns that we should not expect quick, substantial change. Social life is subject to large-scale forces and internal strains that alter it, of course, but because a society is inclined to be in a steady state,[29] the change that occurs in the incidence of society-significant activities tends to be slow and spread over decades. Childbearing outside of marriage provides an example: Non-marital childbearing in the United States underwent a gradual change in incidence over the 1940–2018 period,[30] and its likelihood went down as the educational attainment of the mother and father went up,[31] i.e., as more knowledge was acquired by society and, in turn, by individuals.[32] Both out of-wedlock childbearing and knowledge acquisition are society-significant activities in the United States,[33] but while knowledge acquisition is socially encouraged,[34] out-of-wedlock childbearing is socially disfavored.[35]

In addition to the gradual pace of change in the frequency of key social activities, the amount of such change is likely to be limited, because material change in these

---

Whites in the United States, 1960–1992, 33 DEMOGRAPHY 141, 142 Fig. 1, 146–47, 148 figs. 2 & 3 (1996).

[28] *See* Sharon Sassler & Daniel T. Lichter, *Cohabitation and Marriage: Complexity and Diversity in Union-Formation Patterns*, 82 J. MARRIAGE & FAM. 35, 41–42, 46 (2020) (concluding that cohabitation has not become a substitute for marriage among Americans).

[29] *See* Martin Landau, *On the Use of Functional Analysis in American Political Science*, 35 SOC. RES. 48, 56–57, 64 (1968) (applying structural-functionalism theory to political phenomena; positing that a society, as a "living system," is "self-regulating" and hence stable; contending that a society changes in only its "form and process," not its "material substance"; and postulating that the functions in a system — what the components of a system do to sustain the system — change much more quickly than the structure of the system).

[30] Scc SOLOMON- FEARS, *supra* note 26, at 24–26 tbl. A-1; Martin, *supra* note 26, at 26 tbl. 10.

[31] Kelly Musick, *Planned and Unplanned Childbearing Among Unmarried Women*, 64 J. MARRIAGE & FAM. 915, 919, 923 tbl. 4, 925 (2002); Marcia J. Carlson et al., *Examining the Antecedents of U.S. Nonmarital Fatherhood*, 50 DEMOGRAPHY 1421, 1426, 1438 tbl. 4 (2013). *Accord*, Dawn M. Upchurch et al., *Nonmarital Childbearing: Influences of Education, Marriage, and Fertility*, 39 DEMOGRAPHY 311, 321 tbl. 3 (2002).

[32] SAIL VOL. 2, *supra* note 20, at 46, 119; THE PLACE OF LAW, *supra* note 12, at 308.

[33] *Cf.* SAIL VOL. 1, *supra* note 6, at 81–86 (listing criteria that determine whether a particular social activity is societally important/significant, and suggesting sources that may supply information pertinent to whether a particular social activity satisfies the criteria).

[34] The U.S. Constitution provides that "The Congress shall have Power... To promote the Progress of Science and useful Arts, by securing for limited Times to Authors and Inventors the exclusive Right to their respective Writings and Discoveries." U.S. CONST. art. I, § 8, cl. 8. The foregoing provision, which is known as the Intellectual Property Clause, has provided the authority for federal legislation on copyrights and on patents. Both copyrights and patents are popularly thought to enhance the welfare of society. Eugene R. Quinn, Jr., *An Unconstitutional Patent in Disguise: Did Congress Overstep Its Constitutional Authority in Adopting the Circumvention Prevention Provisions of the Digital Millennium Copyright Act?*, 41 BRANDEIS L.J. 33, 37–43 (2002).

[35] PEW RES. CTR., AS MARRIAGE AND PARENTHOOD DRIFT APART, PUBLIC IS CONCERNED ABOUT SOCIAL IMPACT 49, 50, 51, 73 (2007) (reporting the results of a survey conducted in 2007 of a national sample of adults in the United States; finding that (i) two out of three respondents

frequencies can produce societally unacceptable social instability. A society will not indefinitely tolerate instability beyond a certain point—indeed, a badly destabilized society will probably not wait very long to reduce instability and the societal stress that accompanies it. Because childbearing has long been socially approved and endorsed in the United States[36]—and has thus historically been the chief cause of U.S. population growth (see Fig. 1.4)—law that openly aims to curb the incidence of fertility will encounter political resistance. Given that the content of law reflects patterns of social behavior,[37] law that is expressly and clearly designed to reduce the birth rate in the United States cannot be expected to emerge in the near future, and may not emerge at all.

3.3 Law and Government Policy

If concern with the environment prompts action aimed at curtailing population growth, how effective will the action be? Structural-functionalism theory in sociology implies that law and government policy will not bring about a substantial reduction in population growth, but the theory may, of course, be wrong. To answer the question, therefore, well-designed quantitative research is needed. In this regard, a point to keep in mind is that world-level change in population size is driven entirely by the numerical difference between births and deaths across all countries. Country-specific change in population size, on the other hand, is driven not only by the birth–death numerical difference in the country but also by net international migration to/from the country, i.e., by the difference between the number of immigrants entering the country and the number of emigrants leaving the country. Given the existence of national sovereignties, a response to the question regarding the effectiveness of law and government policy must focus on nations.

The focus here, then, is on the nation-state. To organize the discussion, I distinguish "developed" nations from "developing" nations. The former category is composed of nations that are structurally complex and democratically governed;

---

thought that increased childbearing by "single women... without a male partner to help raise" the children was harmful to society, (ii) three out of five respondents believed that increased childbearing by unmarried couples was hurting society, and (iii) just six percent of the respondents viewed these trends as socially beneficial). *See* Melissa Murray, *What's So New About the New Illegitimacy?*, 20 AM. U.J. GENDER SOC. POL'Y & L. 387, 393, 399, 405, 408–09, 412 (2012) (summarizing, in parts I and II of her article, U.S. Supreme Court decisions relevant to non-marital childbearing; concluding that the Supreme Court, through its decisions on the rights of parents of non-marital children, did not favor out-of-wedlock childbearing but, instead, preferred childbearing within marriage).

[36] *See* GEORGE GAO, PEW RES. CTR., AMERICANS' IDEAL FAMILY SIZE IS SMALLER THAN IT USED TO BE (2015) (reporting that since the mid-1930s, when ideal family size was first measured in U.S. public opinion surveys, at least 90% of American adults in most years have favored a minimum of two children, and sizeable percentages have favored a minimum of three children), https://www.pewresearch.org/fact-tank/2015/05/08/ideal-size-of-the-american-family.

[37] EXPLAINING LAW, *supra* note 6, at 8–11, 47–52.

the latter category is composed of nations that are neither structurally complex nor democratically governed. Although the two categories represent the opposite ends of a continuum and few countries may be at either end, the categories are based on empirical criteria. Having concrete referents, the categories offer a plausible organizing principle for my discussion.

3.3.1 Effectiveness

Let us turn now to what social science research tells us about the degree to which population growth in developed nations may be influenced by law and government policy. I introduce this research with a look at whether the behavior of individuals was altered by law during a prominent historical period in the United States. The period, which occurred in the first half of the twentieth century, is popularly known as Prohibition[38] and involved the adoption of law that sought to promote public health and welfare by eliminating the excessive consumption of intoxicating beverages.[39] The law that was adopted is unusual in that it entailed the addition of an amendment to the national Constitution, not just the passage of a statute.[40] Logically, a constitutional amendment in the United States accompanied by implementing legislation concerned with public wellbeing would be expected to have a large impact on the behavior of individual Americans—the U.S. Constitution is fundamentally a symbol that fosters, and evolves to foster, cohesiveness in American society.[41] The behavioral impact of the law that defined Prohibition is likely, therefore, to signal the maximum behavioral impact that law in a modern nation is able to have on all society-significant social activities.

---

[38]The period has been labelled a "major chapter in American history." Mark Lawrence Schrad, *Constitutional Blemishes: American Alcohol Prohibition and Repeal as Policy Punctuation*, 35 POL'Y STUD. J. 437, 438 (2007). It has also been characterized as "one of the most colorful and controversial" eras in the history of the United States and as involving what was perhaps "the most widely and flagrantly disobeyed" law ever adopted by the U.S. federal government. LOUISE CHIPLEY SLAVICEK, THE PROHIBITION ERA 1 (2009).

[39]SARAH W. TRACY, ALCOHOLISM IN AMERICA: FROM RECONSTRUCTION TO PROHIBITION 8–12, 14–16 (2005).

[40]U.S. CONST. amend. XVIII . The Eighteenth Amendment became part of the Constitution in January 1919 and, by its terms, was effective one year later. THE CONSTITUTION OF THE UNITED STATES OF AMERICA AS AMENDED, H.R. Doc. No. 110–50, at 19–20 (2007) [hereinafter U.S. CONSTITUTION AS AMENDED] (providing the text and ratification record of the Eighteenth Amendment). Federal legislation implementing the Eighteenth Amendment was adopted in October 1919, but was not fully in force until the Amendment took effect in January 1920. National Prohibition Act, Pub. L. No. 66–66, ch. 85, tit. III, § 21, 41 Stat. 305, 322–23 (1919). The Eighteenth Amendment was repealed in December 1933 by the Twenty-First Amendment. U.S. CONSTITUTION AS AMENDED, *supra*, at 22.

[41]Max Lerner, *Constitution and Court as Symbols*, 46 YALE L. J. 1290 (1937); SAIL VOL. 1, *supra* note 6, at 14–15.

A related point should not be overlooked. The values of American citizens are embodied in the provisions of the U.S. Constitution and regulatory statutes that relate to social activities, and prevailing beliefs in the United States about right and wrong are accordingly the cornerstone of U.S. law on activities that American society deems to be public in character.[42] As the U.S. Supreme Court has observed, what law proscribes and prescribes for topics in the public realm "is constantly based on notions of morality."[43] Consequently, a change in the content of regulatory law on social matters expresses a new, presumptive social obligation for individuals. However, the change in law underlying Prohibition produced a sizeable reduction in the consumption of intoxicating beverages only in the short run (one or two years); in the long run, it had no clear impact on consumption.[44] This finding is striking because Americans as a whole could not have been unaware of the new law.

Notably, the failure of Prohibition law to have a large, long-term effect on the consumption of intoxicating beverages is in line with the results of quantitative research on the reaction of numerous types of individual behavior to interventions by government in developed nations: The frequency of society-important social

---

[42]THE PLACE OF LAW, *supra* note 12, at 213–14.

[43]Bowers v. Hardwick, 478 U.S. 186, 196 (1986). *Accord*, Williams v. Pryor, 240 F.3d 944, 949 (11th Cir. 2001) ("The crafting and safeguarding of public morality has long been an established part of the States' plenary police power to legislate..."). *See* 1568 Montgomery Highway, Inc. v. City of Hoover, 45 So.3d 319, 337 (Ala. 2010) (observing that "public morality" is "a rational basis for legislation"); State v. Smith, 766 So.2d 501, 509 (La. 2000) ("commission of what the legislature determines as an immoral act, even if consensual and private, is an injury against society itself").

In *Bowers*, *supra*, the Court had concluded that the liberty guarantee contained in the Fourteenth Amendment to the Constitution was not violated by a state statute that criminalized same-sex sexual activity even when this activity occurred in a private setting. 478 U.S. at 189, 196. In Lawrence v. Texas, 539 U.S. 558, 578 (2003), the Court overruled its holding in *Bowers*. Both cases stemmed from state statutes that allowed criminal penalties to be imposed on persons who engage in homosexual activity inside a residence. 478 U.S. at 187–88; 539 U.S. at 562–63, 566.

Although *Bowers* and *Lawrence* resulted in contrary holdings, the rationales of the Court that underlie these holdings are similar in one respect: They employ, albeit implicitly, the dichotomy between public and private. To be precise, the majority opinion in *Bowers*, by calling attention to the long-standing and widespread statutory bans in the United States on same-sex sexual activity, recognizes that such activity is of concern to the public, regardless of where this activity happens. 478 U.S. at 192–94. The Court in *Bowers* thus assigned homosexual activity to the public sphere. By contrast, the majority opinion in *Lawrence* points out that the statute being challenged "does not involve public conduct," but instead "touch[es] upon the most private human conduct, sexual behavior, and in the most private of places, the home." 539 U.S. at 567, 578. The Court in *Lawrence* accordingly treated homosexual activity that occurs within a setting not visible to the public as being in the private sphere. Mark D. Rosen, *Why the Defense of Marriage Act Is Not (Yet?) Unconstitutional: Lawrence, Full Faith and Credit, and the Many Societal Actors That Determine What the Constitution Requires*, 90 MINN. L. REV. 915, 921 (2006).

[44]Angela K. Dills et al., *The Effect of Alcohol Prohibition on Alcohol Consumption: Evidence from Drunkenness Arrests*, 86 ECON. LETTERS 279, 280, 283 (2005). The time-limited impact of the change in law was probably attributable to the social benefits that accrue from the consumption of intoxicating beverages in group settings. Derek A. Kreager et al., *Delinquency and the Structure of Adolescent Peer Groups*, 49 CRIMINOLOGY 95, 103–05, 109, 119–22 (2011); E. C. Moore, *The Social Value of the Saloon*, 3 AM. J. SOCIOL. 1 (1897). *See also* Bethany L. Peters & Edward Stringham, *No Booze? You May Lose: Why Drinkers Earn More Money than Nondrinkers*, 27 J.

activities in which individual human beings engage is not generally, substantially, and permanently affected by government regulatory action (i.e., law and policy) whose chief purpose is to alter this frequency.[45] The foregoing conclusion, we should keep in mind, is consistent with existing theory—specifically, structural-functionalism theory—and all else being equal, a research-grounded conclusion that a theory subsumes, i.e., an empirical finding that is explained, has greater credibility than a research-grounded conclusion that has no theory to support it. Structural-functionalism theory (see *supra* Part 3.2) is able to explain the common ineffectiveness of socially focused law because it posits that a society, by virtue of being an ordered entity, possesses a proclivity for strong interpersonal ties among its participants (social integration) and for mutually supporting institutions (system integration). These proclivities act to solidify a society and make the society resistant to

---

LAB. RES. 411, 414, 417–19 & tbl. 3, 420 n.7 (2006) (studying a national sample of men and women in the United States who were employed thirty or more hours weekly; finding that men (but not women) who drank at "a bar or tavern" one or more times a month had statistically significant higher earnings than those who did not; and hypothesizing that the increment in earnings was due to the social advantages that men obtained from drinking in groups).

[45] EXPLAINING LAW, *supra* note 6, at 14–15, 55–58; LARRY D. BARNETT, LEGAL CONSTRUCT, SOCIAL CONCEPT 26–36 (1993); Paul H. Robinson & John M. Darley, *Does Criminal Law Deter? A Behavioural Science Investigation*, 24 OXFORD J. LEGAL STUD. 173, 174–75, 197–204 (2004); Helmut Hirtenlehner & Per-Olof H. Wikström, *Experience or Deterrence? Revising an Old but Neglected Issue*, 14 EUR. J. CRIMINOLOGY 485, 486–89, 496–98 (2017).

For a recent study that finds law is ineffective in curbing crime that involves sexuality, see Jeff A. Bouffard & LaQuana N. Askew, *Time-Series Analyses of the Impact of Sex Offender Registration and Notification Law Implementation and Subsequent Modifications on Rates of Sexual Offenses*, 65 CRIME & DELINQ. 1483, 1503–04 (2019) (analyzing monthly data from January 1977 to April 2012 for Harris County, Texas, which includes the city of Houston). The conclusion reached by the Bouffard-Askew study is consistent with the conclusion reached by Jeffrey C. Sandler et al., *Does A Watched Pot Boil? A Time-Series Analysis of New York State's Sex Offender Registration and Notification Law*, 14 PSYCHOL. PUB. POL'Y & L. 284 (2008) (analyzing monthly data covering the state of New York from 1986 through 2006). *See also* Kristin Zgoba et al., *Megan's Law 20 Years Later: An Empirical Analysis and Policy Review*, 45 CRIM. JUST. & BEHAV. 1028, 1029–30, 1033–34, 1044 (2018) (analyzing yearly data covering the state of New Jersey; studying the impact of a state statute that (i) required a convicted sexual offender, after being released from prison and taking up residence in New Jersey, to register with the local police department and (ii) required specified segments of the public to be notified of the presence of the offender; finding that the statute did "not have a demonstrable effect on future offending" in either the short-run or the long run).

A recent study looked at law that requires minors to be in school until a designated age and concluded that this law did not have a uniform impact on school attendance. Erica Raimondi & Loris Vergolini, *Everyone in School: The Effects of Compulsory Schooling Age on Drop-out and Completion Rates*, 54 EUR. J. EDUC. 471, 475, 479–80 tbl. 2, 482–83 (2019) (using time-series data from recurring quarterly sample surveys of households in Italy; analyzing the data to estimate change in the odds of secondary-school attendance and graduation after the law-mandated ending age for compulsory schooling was raised from age 14 to age 15; finding that, net of other independent variables, the higher law-mandated age was responsible for (i) a 38% increase in the odds of school attendance by 16 year-olds, (ii) a 42% increase in the odds of school attendance by 17 year-olds,

social change, including social change emanating from government action that seeks to alter individual behavior.[46]

Let me clarify and delimit the thesis that government does not materially control core forms of social behavior in a society. To do so, I turn to research that has investigated whether national constitutions, in what they say about the physical environment, improve or harm environmental quality. Simply put, the research has found that the presence of environment-safeguarding provisions in the constitutions is associated with better environmental conditions. A study of 168 nations reported that (i) the degree to which the constitution of a nation asserted rights to a healthful environment (i.e., the relative strength of environmental-protection rights in a national constitution) was inversely related to the amount of carbon-dioxide that the nation emitted into the atmosphere; and (ii) the fit of its regression models for CO_2 emissions was excellent ($R^2 \geq 0.73$).[47] Another study reported that a wider range of positive nation-level ecological outcomes was empirically tied to the presence of environment-protecting provisions in the constitutions of the nations, but the fit of the regression models for the index of outcomes was no more than modest ($R^2 < 0.28$).[48] Notably, these studies did not ascertain *how* the environmental benefits were

---

which effect diminished over time and was no longer detectable at the end of the four years (1999–2003) that the new law was in force, and (iii) a 20% decrease in the odds of graduating from secondary school). Earlier research, too, concluded that compulsory-schooling law brought about little overall change in rates of school attendance. THE PLACE OF LAW, *supra* note 12, at 118, 142 n.162 (citing three studies).

[46]The proclivities for social integration and system integration may operate through, inter alia, the feedback loops that exist in a society. *See* Sylvia Walby, *Complexity Theory, Systems Theory, and Multiple Intersecting Social Inequalities*, 37 PHIL. SOC. SCI. 449, 454–55, 463–64 (2007) (conceptualizing societies as systems; pointing out that feedback loops are implicit in the concept of a system; contending that feedback loops can cause the stability of a system to increase or decrease exponentially in response to a stimulus; and citing instances of exponential reductions in economic-system stability but not in social-system stability).

Although feedback loops may produce side effects, the side effects can in turn give rise to reactions that counter or suppress the side effects. *E.g.*, Navid Ghaffarzadegan et al., *Research Workforce Diversity: The Case of Balancing National versus International Postdocs in US Biomedical Research*, 31 SYS. RES. & BEHAV. SCI. 301 (2014). The reactions by a society to side effects can, of course, be expected to generate other side effects that are then followed by societal reactions that neutralize these side effects. Such chains of events would maintain or build society-level inertia, which would curtail the ability of law and government policy to achieve their social goals. Notably, the chains are compatible with and explicable by structural-functionalism theory. *See* Roger I. Roots, *When Laws Backfire: Unintended Consequences of Public Policy*, 47 AM. BEHAV. SCI. 1376, 1380, 1385–88, 1390 (2004) (concluding that law and government policy have rarely cured social ills and have often had side effects; reasoning that this conclusion supports the utility of structural-functionalism theory).

[47]Alessandra Cepparulo et al., *Can Constitutions Bring About Revolutions? How to Enhance Decarbonization*, 93 ENVTL. SCI. & POL'Y 200, 204–05 & tbl. 3 (2019).

[48]Chris Jeffords & Lanse Minkler, *Do Constitutions Matter? The Effects of Constitutional Environmental Rights Provisions on Environmental Outcomes*, 69 KYKLOS 294, 301, 303, 308, 313 n.25, 317 tbl. 6 (2016) (using an index measuring seventeen aspects of "ecosystem vitality").

achieved, but the benefits are likely to have accrued through the use of government-mandated environment-protecting equipment in extracting/processing raw materials and through the installation of government-mandated environment-protecting equipment in manufactured goods. If so, constitutional law yielded environmental benefits by supplying the foundation for government regulation of what *entities* (e.g., business firms) do in the economy, not what individual humans do in social life.[49] Government regulation, in other words, can bring about material, long-term change in targeted entity economic behavior even though it is unable to do so in targeted individual social behavior when that social behavior is of a type that is central to the society.

At least in the United States, childbearing and international migration doubtlessly qualify as society-significant activities,[50] but neither of these activities appears to exhibit a large, lasting response to government regulation that is aimed at them. To be precise, a credible body of social science evidence indicates that, in developed nations, government is unable to have a sizeable, long-term effect on the incidence

---

[49]Both studies have several limitations that should be mentioned because the findings of the studies may have been affected by one or more of the limitations.

First, neither of the studies specifically controlled between-nation differences in culture. Thus, neither of the studies removed the sizeable influence that culture normally exercises over the content and operation of law. For quantitative research on this influence, see Amir N. Licht et al., *Culture Rules: The Foundations of the Rule of Law and Other Norms of Governance*, 35 J. COMP. ECON. 659, 667 (2007); Daniel Kaufmann et al., *Governance Matters III: Governance Indicators for 1996–2002*, at 2–4 (World Bank Pol'y Res. Working Paper No. 3106, 2003) (describing the indexes comprising the dependent variables in the study by Licht et al.). *Cf.* Fikresus Fikrejesus Amahazion, *Human Rights and World Culture: The Diffusion of Legislation Against the Organ Trade*, 36 SOCIOL. SPECTRUM 158 (2016) (positing a geographic spread of cultural beliefs that individuals have intrinsic worth and finding that empirical indicators of these beliefs predicted the adoption by nations, during the 1965–2012 period, of law prohibiting commercial transactions in human organs). For qualitative research on the influence that culture has on law, see, for example, Peter Just, *Let the Evidence Fit the Crime: Evidence, Law, and "Sociological Truth" among the Dou Donggo*, 13 AM. ETHNOLOGIST 43, 58–59 (1986).

The second limitation of the studies is that they each employed data that were in essence cross-sectional, not longitudinal. Cepparulo et al., *supra* note 47, at 204; Jeffords & Minkler, *supra* note 48, at 308–10. Since a cause-effect relationship involves a *sequence* of events, such a relationship must be proven using longitudinal data. A relationship between variables that appears in cross-sectional data may be, and often is, absent in longitudinal data. FRANS L. LEEUW & HANS SCHMEETS, EMPIRICAL LEGAL RESEARCH 122–25 (2016).

The third limitation of the studies is that they did not attempt to ascertain whether one or more nations in their data were influential outliers. As a result, neither study in its statistical analysis took into account nations that may have disproportionately shaped the relationships found between the independent variables and dependent variable. Although the procedures for identifying and treating influential outliers are not uniform due to a lack of generally accepted guidelines, outliers require attention because they can distort the results of regression analyses. Herman Aguinis et al., *Best-Practice Recommendations for Defining, Identifying, and Handling Outliers*, 16 ORG. RES. METHODS 270, 271, 273, 297 (2013).

[50]Text accompanying *supra* note 25; *supra* note 36; *supra* Chap. 1, at 14–18. The U.S. Constitution, in authorizing Congress "[t]o establish an [sic] uniform Rule of Naturalization," implicitly recognizes immigration by noncitizens because naturalization presupposes such immigration. U.S. CONST. art. 1, § 8, cl. 4.

of childbearing[51] and that, in the United States, government has not had much of a long-run impact on the incidence of migration into the country.[52] The finding that change in the content of government regulatory action in a developed nation may temporarily exert a material influence on the incidence of a targeted society-central social activity should be tempered, accordingly, by the recognition that it does not

---

[51]See the following summaries of empirical research: Anne H. Gauthier, *The Impact of Family Policies on Fertility in Industrialized Countries: A Review of the Literature*, 26 POPULATION RES. & POL'Y REV. 323, 342 (2007) (concluding that, in developed nations, fertility-relevant policies have only a "small" effect on the frequency of childbearing); Olivier Thévenon & Anne H. Gauthier, *Family Policies in Developed Countries: A "Fertility-Booster" with Side-Effects*, 14 COMMUNITY, WORK & FAM. 197, 211 (2011) (concluding that "family-friendly policies" in developed nations have just a "limited" effect on the fertility rates in these nations). *Accord*, Ann-Zofie Duvander et al., *Impact of a Reform Towards Shared Parental Leave on Continued Fertility in Norway and Sweden*, POPULATION RES. & POL'Y REV. (2020), https://doi.org/10.1007/s11113-020-09574-y; Suzanne Ryan et al., *State-level Welfare Policies and Nonmarital Subsequent Childbearing*, 25 POPULATION RES. & POL'Y REV. 103 (2006). *See also* Abdoulaye Quedraogo et al., *Fertility and Population Policy*, 42 PUB. SECTOR ECON. 21, 22, 36–37 (2018) (using panel data on 133 nations for the years 1976–2013; finding that the governments of these nations did not affect the level of childbearing through policies that were adopted to increase fertility, e.g., policies that provided tax inducements for childbearing and monetary assistance to defray childrearing and housing costs; also finding that the governments decreased childbearing through policies that were not intended to reduce fertility, e.g., policies that improved education, health care, and family-planning services); Yeon Jeong Son, *Do Childbirth Grants Increase the Fertility Rate? Policy Impacts in South Korea*, 16 REV. ECON. HOUSEHOLDS 713, 716, 733 (2018) (using panel data on 230 municipalities in South Korea for the years 2001–2014; finding that municipalities that adopted a policy of providing monetary grants to parents for the birth and care of children experienced an increase in childbearing; but concluding that, to raise the level of childbearing substantially, the amount of the grants would have to be very large and that such an amount would not be "cost-effective").

One research summary maintains that a set of family-supportive policies can increase fertility when the policies embody "a spirit" that favors childbearing and exist within "a family-friendly culture." Jan M. Hoem, *Overview Chapter 8: The Impact of Public Policies on European Fertility*, 19 DEMOGRAPHIC RES. 249, 253–56 (2008). The degree to which these policies are effective, in other words, is determined by their societal context. *Id.* at 253–55. That context, however, is what gives rise to the content of law and government policy on society-significant social behaviors in developed nations. The assertion by Dr. Hoem, therefore, is not inconsistent with the thesis that societal culture and structure, not law and government policy, are the chief determinants of the level of childbearing in developed countries. EXPLAINING LAW, *supra* note 6, at 8–9, 17, 50–52.

[52]Wayne A. Cornelius, *Impacts of the 1986 US Immigration Law on Emigration from Rural Mexican Sending Communities*, 15 POPULATION & DEV. REV. 689, 701–02 (1989); Douglas S. Massey et al.,

do so in the long run.[53] Indeed, change in law content in a developed nation seems to be an outcome, rather than a cause, of change in the incidence of the activities.[54]

A final point should be made. In a developing nation, law and government policy may or may not exert a large, enduring influence on how many or how often individuals participate in society-important social activities. Unfortunately, we simply do not know, because the evidence that presently exists is insufficient to establish the long-term effectiveness of action taken by the government of a developing nation when that action expressly targets a society-central social activity and attempts to alter its frequency among individuals. Nonetheless, if we proceed on a "best guess" basis, a reasonable conclusion from the available evidence is that, in both developed and developing nations, government does not have a sizeable, permanent impact on the frequency of key forms of individual-level social behavior when its principal intention is to affect that frequency.

3.3.2 Collateral Consequences

Although structural-functionalism theory and empirical research offer no reason to believe that law expressly intended to curb population growth will achieve its goal, the current state of knowledge about the societal function of law is incomplete and contains considerable uncertainty. We thus turn to the question of whether environmental law scholarship must simply accept Malthusian solutions (e.g., high-mortality epidemics[55]) to the population problem or whether environmental law

---

Why Border Enforcement Backfired, 121 AM. J. SOCIOL. 1557, 1558 (2016) (summary of research); Pia M. Orrenius & Madeline Zavodny, *Do Amnesty Programs Reduce Undocumented Immigration? Evidence from IRCA*, 40 DEMOGRAPHY 437, 448 (2003); Emily Ryo, *Deciding to Cross: Norms and Economics of Unauthorized Migration*, 78 AM. SOCIOL. REV. 574, 580–81, 588–89 tbl. 2, 593 (2013). Intensified actions by the U.S. government to control Mexico-to-U.S. migration between 1970 and 2010 evidently enlarged the number of U.S. residents from Mexico who were illegally present in the United States. Massey et al., *supra*, at 1590–91.

[53]Short-term effects that law has on the frequency of society-important social behavior must be juxtaposed with the absence of long-term effects of the same law. Macrosociology needs to go beyond simply noting the difference in impact between the former and the latter; it should be able to explain the disappearance of an effect over time.

[54]EXPLAINING LAW, *supra* note 6, at 47–52.

[55]Infectious diseases have not only killed large numbers of humans in the past but have the potential to do so in the future. R. M. Anderson & R. M. May, *The Invasion, Persistence and Spread of Infectious Diseases Within Animal and Plant Communities*, 314 PHIL. TRANSACTIONS OF THE ROYAL SOC'Y OF LONDON SERIES B 533, 533–34 (1986); Alexander Koch et al., *Earth System Impacts of the European Arrival and Great Dying in the Americas after 1492*, 207 QUATERNARY SCI. REV. 13, 22 tbl. 4, 30 (2019); Michael T. Osterholm, Univ. of Minnesota Ctr. for Infectious Disease Research & Pol'y, *Commentary: Pandemic Preparedness and Missed Opportunities* (2017) (contending that "pandemic clocks are ticking; we just don't know what time it is"), https://www.cidrap.umn.edu/news-perspective/2017/10/commentary-pandemic-preparedness-and-missed-opportunities (last visited March 18, 2020). *See also* notes 59 & 91 and accompanying text in *supra* Chap. 1; Katherine F. Smith et al., *Global Rise in Human Infectious Disease*

scholarship is able to propose concrete steps that will be effective in substantially diminishing growth in the size of the human population. Effective steps may not be possible, because a material reduction in population growth faces strong headwinds.[56] Nonetheless, assuming that Malthusian solutions are avoidable,[57] I believe that law and government policy may help to control population growth. In particular, law and government policy whose sole or primary expressed goal is *other* than a reduction in fertility or immigration may bring about a substantial, lasting reduction in fertility or immigration even while law and government policy that is aimed explicitly at fertility or immigration enjoys no more than temporary success.[58] Put succinctly, the instruments of government may produce an outcome that they do not and cannot seek directly.

To illustrate, the incidence of childbearing is materially limited by the amount of space in residential housing.[59] Policies that promote higher human density in

---

Outbreaks, 11(101) J. OF THE ROYAL SOC'Y INTERFACE, Dec. 2014, at 1, 5 (finding that among humans the worldwide number of outbreaks of infectious disease, and the number of new infectious diseases, increased between the period 1980–1984 and the period 2005–2010), https://royalsocietypublishing.org/doi/pdf/10.1098/rsif.2014.0950.

[56] Patrick Gerland et al., *World Population Stabilization Unlikely This Century*, 346 SCI. 234, 234 (2014) (modeling world population growth; forecasting that a large numerical increase will occur in the size of the human population during the twenty-first century unless childbearing decreases significantly in sub-Saharan Africa, and stating that such decreases would be "unprecedented").

[57] According to the widely discussed study *The Limits to Growth*, which was published in the early 1970s and used computer simulation to estimate the long-term consequences for the human species of trends in resource consumption and environmental contamination, Malthusian solutions will occur during the twenty-first century unless there is substantial change in prevailing forms of human activity. Jorgen Randers, *The Real Message of* THE LIMITS TO GROWTH, 21 GAIA 102 (2012) (summarizing *The Limits to Growth* study and noting its ramifications). This conclusion is buttressed by recent evidence. Graham M. Turner, *On the Cusp of Global Collapse? Updated Comparison of* THE LIMITS TO GROWTH*with Historical Data*, 21 GAIA 116, 123 (2012) (adding forty years of data to three simulations in *The Limits to Growth* study, which employed data covering the 1900–1970 period; finding that inclusion of the added data provided support to the simulation in which patterns of social and economic behavior persist; and pointing out that this simulation predicts a "collapse of the global economy and population in the near future").

For computer-simulation evidence that national policies that avert Malthusian solutions are feasible, see Surya Raj Acharya & Khalid Saeed, *An Attempt to Operationalize the Recommendations of the "Limits to Growth" Study to Sustain the Future of Mankind*, 12 SYSTEMS DYNAMICS REV. 281 (1996).

[58] *E.g.*, Quedraogo et al., *supra* note 51.

[59] Hill Kulu & Andres Vikat, *Fertility Differences by Housing Type: The Effect of Housing Conditions or of Selective Moves?*, 17 DEMOGRAPHIC RES. 775, 779–80, 790 (2007); Hill Kulu et al., *Settlement Size and Fertility in the Nordic Countries*, 61 POPULATION STUD. 265, 270–71, 278 (2007); Nathanael Lauster, *A Room to Grow: The Residential Density-Dependence of Childbearing in Europe and the United States*, 37 CANADIAN STUD. IN POPULATION 475, 481, 485, 487, 488 tbl. 3, 491 (2010). *See also* George Martine et al., *Urbanization and Fertility Decline: Cashing in on Structural Change* 19, 25, 28–30 & tbl. 5 panel B (Internat'l Inst. for Env't & Dev., IIED Working Paper, 2013) (analyzing cross-sectional data for 131 countries around the year 2010; finding that,

housing, therefore, are likely to help curtail the birth rate,[60] although the speed with which they will do so in the future is the subject of disagreement.[61] These policies, however, not only have the potential to curb population growth and its ecological impacts[62]; they may also be able to curtail the per-capita consumption of energy.[63] The reduction in energy consumption warrants mention because governments have started to reexamine how they plan urban areas and to recognize the energy-use

all else being equal, the percentage of the population of a country that resided in an urban area was inversely related to the total fertility rate in the country).

[60]In the United States, urban-containment policies of local governments and growth-management policies of state governments have been found to (i) reduce the number of square miles of land necessary to accommodate the resident population of an urbanized area — a reduction that raises population density — as well as (ii) curtail the mean number of individuals in a household. Robert W. Wassmer, *The Influence of Local Urban Containment Policies and Statewide Growth Management on the Size of United States Urban Areas*, 46 J. REGIONAL SCI. 25, 34, 41–42 tbl. 3 (2006) (studying 452 places in the United States that in the year 2000 were designated urbanized areas by the U.S. Census Bureau).

[61]If policies that increase population density in housing lower the birth rate slowly, they may not substantially alter the course of world population growth during the twenty-first century and yield large environmental benefits. *Compare* Corey J. A. Bradshaw & Barry W. Brook, *Human Population Reduction is not A Quick Fix for Environmental Problems*, 111 PROC. NAT'L ACAD. SCI. 16,610, 16,613–15 (2014) (estimating that growth in world population during the twenty-first century will be material and arguing that this growth is "virtually locked-in"), *with* Brian C. O'Neill et al., *Global Demographic Trends and Future Carbon Emissions*, 107 PROC. NAT'L ACAD. SCI. 17,521, 17,525 (2010), *and* Brian C. O'Neill et al., *Plausible Reductions in Future Population Growth and Implications for the Environment*, 112 PROC. NAT'L ACAD. SCI. E506 (2015) (critiquing the Bradshaw-Brook estimates and contending that world birth rates can decline sufficiently during the twenty-first century to bring about "significant" reductions in the production of greenhouse gases).

[62]These impacts are summarized in Part 1.4 of *supra* Chapter 1. *See also* Ahmet Atıl Aşıcı & Sevil Acar, *Does Income Growth Relocate Ecological Footprint?*, 61 ECOLOGICAL INDICATORS 707, 712 tbl. 2 (2016) (studying 116 nations at different income levels during the years 2004–2008; finding that increases in population density reduce human-induced environmental impacts stemming from domestic production activities but have no effect on environmental impacts stemming from imports; reporting that the overall R^2 for the relationship between population density and production-generated environmental impacts is just .07 (model (2)).

[63]Brian E. Green, *Explaining Cross-National Variation in Energy Consumption: The Effects of Development, Ecology, Politics, Technology, and Region*, 34 INT'L J. SOCIOL. 9, 20–21 tbl. 2, 24–27, tbl. 4 (models 3, 4, and 5) (2004). *See* Burak Güneralp et al., *Global Scenarios of Urban Density and Its Impacts on Building Energy Use Through 2050*, 114 PROC. NAT'L ACAD. SCI. EARLY ED.

benefit of high density housing.[64] The prospect of diminished energy use can provide an impetus for public policies that explicitly encourage or require high density in housing and that, to the extent they increase this density, cut back on fertility as a source of human population growth.[65]

3.4 Cumulative Birth Rate and Total Fertility Rate

Every law/policy, regardless of the topic addressed, should be assessed to ascertain whether it is succeeding or failing. An assessment of law/policy that may have an effect on the rate at which humans procreate unavoidably involves two measures of childbearing that demographers have developed, and I therefore consider the measures here. The measures are the Cumulative Birth Rate ("CBR") and the Total Fertility Rate ("TFR"). As will be evident, a specific numerical value of the CBR will be set as the goal of childbearing-pertinent law/policy, and the TFR will reveal whether the level of childbearing is moving toward the goal or away from it. The TFR is able to show the direction of the CBR because, as yearly birth rates change, the numerical value of the TFR responds more quickly than the numerical value of the CBR.

An understanding of the CBR and the TFR can begin with their similarities. Specifically, both the CBR and the TFR are calculated in terms of females rather than males; are expressed as the number of children born alive per 1,000 females or per female; and vary across calendar years. What differentiates the two measures is their temporal orientation — the CBR captures procreation to date and thus looks backward in time; the TFR is a forecast of future procreation and thus looks ahead

8945, 8947–48 (2017) (estimating the impact of urban-population density in regions of the world on energy use in each of these regions during the period 2010–2050).

[64] ORG. FOR ECON. CO-OPERATION AND DEV. & CHINA DEV. RES. FOUND., TRENDS IN URBANISATION AND URBAN POLICIES IN OECD COUNTRIES: WHAT LESSONS FOR CHINA? 78, 145, 148 (2010), *available at* www.oecd.org/urban/roundtable/45159707.pdf. *See generally* Nicole Stelle Garnett, *Planning for Density: Promises, Perils and a Paradox*, 33 J. LAND USE & ENVTL. L.1 (2017) (analyzing the policy push for urban redevelopment, and hence for higher housing density, in the United States).

[65] Among individuals, parenthood has been hypothesized to create concern with the environment, because parents are assumed to care about the legacy they leave their children. The hypothesis, however, lacks firm empirical support. Gregory O. Thomas et al., *The Impact of Parenthood on Environmental Attitudes and Behaviour: A Longitudinal Investigation of the Legacy Hypothesis*, 39 POPULATION & ENV'T 261, 262–63, 265–68, 272 (2018) (finding that the birth of a child did not have a general, material influence on the environment-pertinent views and actions of the parents of the child). A collective reduction in childbearing, therefore, is unlikely to alter the prevalence of views and actions bearing on the environment.

in time.[66] Unfortunately, the information that the CBR supplies is frequently and incorrectly thought to be supplied by the TFR.

Given that the two measures will be central to judging childbearing-focused population policy, let me delve more fully into them. I begin with the TFR because of the confusion that often attends it. The TFR is the mean number of children that females in a particular cohort will bear *if* they survive until the end of their reproductive period *and*, while going through all of the ages within that period, have at every age a rate of giving birth that exactly matches the birth rate that characterizes the age during the calendar year for which the TFR is computed. The TFR is accordingly described as the number of live births that

> females would have if they experienced throughout their childbearing years the age-specific birth rates observed in a given year. It can be interpreted as the average number of lifetime births women may be expected to have if they bore children at the rates that women of all ages did in that given year.[67]

In effect, then, the TFR assumes that the average female in a cohort fast-forwards through her entire reproductive period in a single year, viz., the year when the TFR is calculated. The TFR, therefore, is a gauge of hypothetical, not actual, childbearing.

The focus of the CBR contrasts sharply with the focus of the TFR. In appraising the impact of law/policy on childbearing, the CBR is of primary importance because it measures the number of births that women in a cohort had during the years when they were physically able to conceive and carry a pregnancy to term. That time span is usually assumed (and will be assumed here) to end on the fiftieth birthday of women.[68] Both the CBR and the TBR employ this assumption. However, while the CBR counts total childbearing by women before age 50, the TFR forecasts it. Notably, the CBR for a cohort can deviate substantially from the level of childbearing predicted by the TFR for that cohort in the year in which the members of the cohort were born. Actual childbearing, in other words, does not necessarily correspond to projected childbearing. Additionally, the numerical value of the CBR in a given calendar year can deviate from the numerical value of the TFR in the same year. Figure 3.1, which uses data for the United States, illustrates the foregoing points. Both the CBR and the TFR are expressed in the figure as the number of births per woman, and include women of all races.[69]

---

[66] NAT'L CTR. FOR HEALTH STAT., TECHNICAL APPENDIX TO THE COHORT FERTILITY TABLES FOR ALL, WHITE, AND BLACK WOMEN: UNITED STATES, 1960- 2005, at 11 (2010), https://www.cdc.gov/nchs/nvss/cohort_fertility_tables.htm.

[67] U.S. DEPT. OF HEALTH & HUMAN SERV., TRENDS IN THE WELL- BEING OF AMERICA'S CHILDREN AND YOUTH, 2002, at 28 (2002), https://aspe.hhs.gov/report/trends-well-being-americas-children-and-youth-2002.

[68] Among women who are 50 years of age or older, childbearing is rare but not nonexistent. In 2018, to give an example, U.S. women who were 50–54 years of age (the oldest age group for which birth data are reported) had fewer than 1,000 births; the total number of births in 2018 to all U.S. women was nearly 3.8 million. Joyce A. Martin et al., *Births: Final Data for 2018*, NAT'L VITAL STAT. REP., vol. 68, no. 13 (Nov. 27, 2019), at 4, 14 tbl. 3.

[69] The data for the CBR were obtained from: NAT'L CTR. FOR HEALTH STAT., FERTILITY TABLES FOR BIRTH COHORTS BY COLOR: UNITED STATES, 1917–73, at 125 tbl. 5-A (1976) (data for 1940

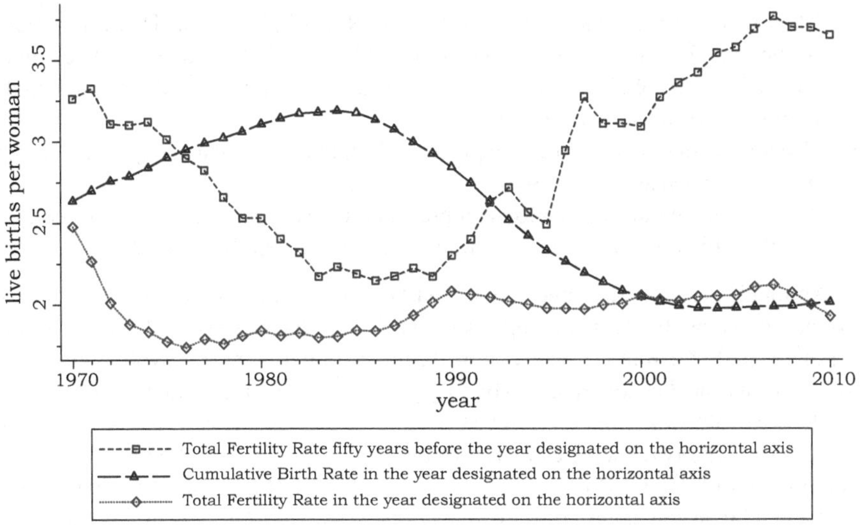

Fig. 3.1 Cumulative Birth Rate and Total Fertility Rate: United States. Source: See footnote 69 in this chapter

Figure 3.1, as a reader will observe, contains three sets of line-connected markers that start with calendar year 1970 and stop with calendar year 2010. The markers in each set are for yearly birth cohorts of women, i.e., cohorts that are defined by the calendar year in which the women in the cohort were born. The square-shaped markers show, for the calendar year designated on the horizontal axis of the figure, the TFR from fifty years earlier, i.e., the TFR in the birth year of the cohort. The triangle-shaped markers show, for the calendar year specified on the horizontal axis, the CBR of each birth cohort of women. The diamond-shaped markers show the TFR for the then-current calendar year, i.e., for the same calendar year as the CBR.

An illustration may help to clarify Fig. 3.1. The illustration employs calendar year 1970, the initial year on the horizontal axis of the figure.

---

to 1960) [hereinafter FERTILITY TABLES]; Nat'l Ctr. for Health Stat., Cohort Fertility Tables – Table 2. Cumulative Birth Rates, by Live-Birth Order, Exact Age, and Race of Women in Each Cohort from 1911 through 1991: United States, 1961–2006 (data for 1961 to 2006); Nat'l Ctr. for Health Stat., Cohort Fertility Tables – Table 2. Cumulative Birth Rates, by Live-Birth Order, Exact Age, and Race of Women in Each Cohort from 1957 through 1995: United States, 2007–2010 (data for 2007 to 2010). For the CBR, the age of women in a given year is their age on January 1 of that year. The documents cited in this footnote are available at https://www.cdc.gov/nchs/nvss/cohort_fertility_tables.htm (last visited Oct. 1, 2020).

The data for the TFR were obtained from: FERTILITY TABLES, *supra*, at 4 tbl. 1A (data for 1920 through 1939); Nat'l Ctr. for Health Stat., *Table 1–7: Total Fertility Rates and Birth Rates, by Age of Mother and Race: United States, 1940–2003*, VITAL STATISTICS OF THE UNITED STATES, 2003, VOL. I, NATALITY (data for 1940 to 1969), https://www.cdc.gov/nchs/products/vsus/vsus_1980_2003.htm (last visited Oct. 1, 2020); Joyce A. Martin et al., *Births: Final Data for 2015*, NAT'L VITAL STAT. REP., vol. 66, no. 1 (Jan. 5, 2017), at 20 tbl. 4 (data for 1970 to 2010).

- The top marker (a square) represents the TFR in 1920, i.e., the TFR for females in the United States who were born in 1920. The TFR for these females, who were fifty years old in 1970, is based on the age-specific birth rates of the female population of the country in 1920.
- The middle marker (a triangle) represents the CBR of women in the United States who were 50 years old in 1970.
- The bottom marker (a diamond) represents the TFR for the 1970 birth cohort, i.e., the TFR for females in the United States who were born in 1970.

Although Fig. 3.1 has three sets of markers, its purpose is to let readers make two comparisons. In the first comparison, the TFR in the birth year of each cohort is compared to the CBR of the same cohort. The first comparison thus involves measures for points in time that are separated by fifty years. In the second comparison, the CBR of fifty-year-old women in a given year is compared to the TFR of the women born in that year. The two exercises therefore involve comparing (1) the birth-year TFR of a cohort with the CBR of that cohort and (2) the CBR at a specific point in time with the TFR at the same point in time. In the case of 1970, the first comparison involves the top marker (the TFR in 1920) in relation to the middle marker (the CBR in 1970), while the second comparison involves the middle marker (the CBR in 1970) in relation to the bottom marker (the TFR of women born in 1970). Comparisons, of course, should be made not just for one calendar year but for many. Doing so reveals change across time.

What conclusions can be drawn from these comparisons? First, childbearing in the United States during the time span covered by the figure was evidently affected materially by important alterations in American social life. This conclusion is suggested by the divergent movements of (i) the TFR of the birth cohorts and (ii) the CBR for these birth cohorts. As Fig. 3.1 shows, the level of childbearing forecasted for the cohorts in the years of their birth changed over time in a manner that approximated a U while the actual childbearing of these cohorts changed in a manner that approximated an ∩ . The markedly different paths of the birth-cohort TFR and the birth-cohort CBR are striking, and likely resulted from substantial, relatively rapid shifts in social conditions. Since in developed nations such shifts will probably occur in most time periods that cover several decades, projected levels of childbearing in these nations will often deviate by a wide margin from actual levels of childbearing. The TFR, in other words, is not, and should not be equated to, the CBR.

The second conclusion from Fig. 3.1 is that the course of the CBR across birth cohorts was predicted by the level and path of the contemporaneous TFR of the cohorts. The figure shows that the CBR from 1970 to the mid-1980s was eventually altered by the much-lower co-occurrent TFR. As a consequence, the CBR after the mid-1980s moved noticeably toward, and by 2000 resembled, the same-year TFR.

As a corollary to the above conclusions, readers should keep in mind that the numerical values of the TFR and CBR do not alone determine whether the number of people in a population will increase, decrease, or stabilize. Other demographic aspects of the population play a role, too. In exploring this point, I will assume that a population characterized by a steady numerical size is desired.

Although a TFR equal to 2.1 is widely believed to be "replacement-level fertility"—i.e., childbearing at a rate that leads a population to plateau in size—the level of childbearing that does so will vary with, inter alia, the mortality rate in the population. Not surprisingly, the childbearing level that produces a numerically constant population will need to be lower when the mortality rate of the population is lower.[70] The level of childbearing required for a constant population size also depends on the average age at which women in the population procreate. For example, an average childbearing age of 25.0 yields four generations per century while an average childbearing age of 33.3 yields three generations. As a result, the fertility level that stabilizes the numerical size of a population can be higher when the women in the population have a higher average childbearing age. Additionally, the mix of demographic factors that generates a numerically flat population includes immigration and emigration. In covering migrants, however, the mix must not be confined to the number of people moving into a population and the number of people moving out of it. The age distribution and the sex ratio of the migrants must be considered as well, because both the age distribution and sex ratio affect rates of childbearing and mortality in a population.

In sum, population size is determined by multiple demographic variables, and all of these variables must be taken into account when predicting how many people may be in a population at a given point in time. Making such predictions, however, will entail modeling statistically the simultaneous operation of the variables and estimating their collective impact on population size. Evaluations of government action that could affect the size of the human population will, therefore, not be simple. Demographers will unavoidably be involved, and participants in the evaluations who are not demographers will need to understand concepts in demography such as the Cumulative Birth Rate and the Total Fertility Rate.

3.5 Closing Comments

"We have met the enemy and he is us."[71]

In conclusion, compelling reasons exist for environmentalists to concern themselves with the size and growth of the human population. Put simply, the number of human beings already on Earth is harming the ecosystem of the planet,[72] and this number

[70]Thomas J. Espenshade et al., *The Surprising Global Variation in Replacement Fertility*, 22 POPULATION RES. & POL'Y REV. 575, 577, 580–81 & tbl. 1 (2003).

[71]WALT KELLY, POGO: WE HAVE MET THE ENEMY AND HE IS US (1972).

[72]INTERGOVERNMENTAL SCIENCE-POLICY PLATFORM ON BIODIVERSITY AND ECOSYSTEM SERVICES, THE GLOBAL ASSESSMENT REPORT ON BIODIVERSITY AND ECOSYSTEM SERVICES: SUMMARY FOR POLICYMAKERS 12, 13, 15, 29, 37 (2019) [hereinafter IPBES REPORT]; Paul R. Ehrlich & Anne H. Ehrlich, *Can A Collapse of Global Civilization Be Avoided?*, 280 PROC. ROYAL SOC'Y B: BIOLOGICAL SCI. 1, 6 (no. 1754, 2013) (summarizing estimates of the capacity of Earth to sustain the current size of the human population given (i) existing technology and (ii) present

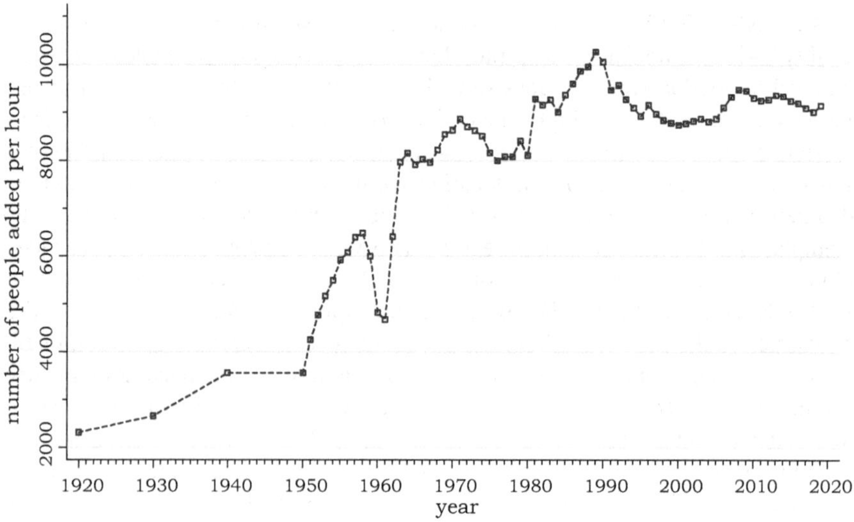

Fig. 3.2 Hourly Increase in World Population (1920–2019). Source: See footnote 73 in this chapter

is rapidly climbing—an average of roughly 9,150 people were added *hourly* to the population of the world during the most recent completed calendar year (2019). As Fig. 3.2 shows, the 2019 hourly addition, while below the maximum reached during the covered period, has not experienced a sustained decline.[73] Instead, it has varied within a bounded, relatively limited range of hourly increases since at least the early 1980s and, arguably, since the mid-1960s. The numbers within this range, furthermore, are large in absolute terms. Earth, therefore, is and has been undergoing a steady and substantial rise in the number of its human inhabitants—a point that Fig. 1.1 in Chapter 1 also makes.

The increasing count of human beings and the resulting greater environmental damage inflicted by the human species should not be merely shrugged off. Although *Homo sapiens* has been altering the biosphere for millennia,[74] the size and activities of its population in recent times have probably brought about a new geological epoch,

---

types of resource use and present levels of resource consumption; observing that the estimates indicate that this capacity has been exceeded; and concluding that the probability of averting an uncontrollable breakdown in world civilization during the twenty-first century is evidently "small"). In addition, see Part 1.4 in *supra* Chap. 1.

[73]Figure 3.2 is based on the data that were used to construct Fig. 1.1. See note 27 in *supra* Chap. 1 for the sources of the data. My calculations for Fig. 3.2 assumed that each year had 8,766 hours, a number obtained by spreading the extra day in a leap year over four years, i.e., by imputing 365.25 days to every calendar year.

[74]Nicole L. Boivin et al., *Ecological Consequences of Human Niche Construction: Examining Long-term Anthropogenic Shaping of Global Species Distributions*, 113 PROC. NAT'L ACAD. SCI. 6388 (2016). In addition, see note 69 in *supra* Chap. 1.

viz., the Anthropocene.[75] Entry into the Anthropocene epoch is significant because it means that, in the words of an aphorism, "the future is not what it used to be."[76] Indeed, *Homo sapiens* has set in motion a planet-wide experiment whose outcome, while not certain,[77] is likely to be undesirable at best, and may even be catastrophic.[78] Notably, disasters worldwide underwent a "sharp increase" in frequency between 1980–1999 and 2000–2019.[79] Further increases may, of course, occur. Moreover, a survey conducted in 2014 of scientists who were members of the American Association for the Advancement of Science found that four out of five respondents believed

---

[75]Scholars have not yet agreed on a single definition of the word "Anthropocene." Valentí Rull, *The "Anthropocene": Neglects, Misconceptions, and Possible Futures*, 18 SCI. & SOC'Y 1056 (2017). Nonetheless, the contention that the Earth has entered another geological epoch is accepted by many scholars. *E.g.*, Jan Zalasiewicz et al., *When Did the Anthropocene Begin? A Mid-Twentieth Century Boundary Level is Stratigraphically Optimal*, 383 QUATERNARY INT'L 196, 201 (2015) (contending that Earth entered the Anthropocene epoch in the middle of the twentieth century when industrialization accelerated and started to have global, synchronized effects).

[76]The aphorism was evidently first put into writing in 1937 and is credited to three literary figures, viz., Laura Riding, Robert Graves, and Paul Valéry. Quote Investigator, https://quoteinvestigator.com/2012/12/06/future-not-used (last visited July 14, 2020).

[77]Jane Lubchenco, *Entering the Century of the Environment: A New Social Contract for Science*, 279 SCI. 491, 492 (1998).

[78]Jane Lubchenco et al., *The Sustainable Biosphere Initiative: An Ecological Research Agenda: A Report from the Ecological Society of America*, 72 ECOLOGY 371, 377 (1991); LAURIE LAYBOURN-LANGTON ET AL., INST. PUB. POL'Y RES., THIS IS A CRISIS: FACING UP TO THE AGE OF ENVIRONMENTAL BREAKDOWN 4–5, 14–15, 20, 31 (2019), https://www.ippr.org/research/publications/age-of-environmental-breakdown; Digby J. McLaren, *Population Growth — Should We Be Worried?*, 17 POPULATION & ENV'T 243, 255, 257–58 (1996); Randers, *supra* note 57; Turner, *supra* note 57. *See also* IPBES REPORT, *supra* note 72, at 12, 15, 37 (observing that "[t]he rate of global change in nature during the past 50 years is unprecedented in human history" and naming, inter alia, population growth as a cause of this change).

If the ecosystem of the Earth has thresholds that when passed trigger severe global environmental effects, the consequences of exceeding the thresholds would be world-wide and potentially cataclysmic. The natural sciences, however, are unsure that a broad range of such thresholds exist. *Compare* Barry W. Brook et al., *Does the Terrestrial Biosphere Have Planetary Tipping Points?*, 28 TRENDS IN ECOLOGY & EVOLUTION 396 (2013) (concluding that presently available empirical evidence indicates that there are few major ecological thresholds), *with* José A. Rial et al., *Nonlinearities, Feedbacks and Critical Thresholds within the Earth's Climate System*, 65 CLIMATIC CHANGE 11 (2004) (describing the sources of tipping points in the climate of the planet). *See also* Yiqi Luo et al., *Coordinated Approaches to Quantify Long-Term Ecosystem Dynamics in Response to Global Change*, 17 GLOBAL CHANGE BIOLOGY 843, 846 tbl. 1, 852 (2011) (pointing out and discussing the obstacles encountered in empirical research on ecosystem thresholds).

[79]CTR. FOR RES. ON THE EPIDEMIOLOGY OF DISASTERS, U.N. OFFICE FOR DISASTER RISK REDUCTION, HUMAN COST OF DISASTERS: AN OVERVIEW OF THE LAST 20 YEARS, 2000–2019, at 6 ([2020]). For more information on this report, see note 96 in *supra* Chapter 1.

that world population growth will lead to shortages of food and natural resources.[80] Even without such shortages, however, markedly higher rates of human mortality could well occur if bacteria become increasingly resistant to antibiotics[81]: Growth in the number of people furthers such resistance through increased human population density[82] and through more-frequent human activities that promote climate warming.[83] Human death rates could escalate, too, if there are pandemics involving viruses that are carried by wildlife, that infect humans who come into contact with the wildlife, and that, once in humans, are transmitted person to person.[84] A consequence of a larger human population is, of course, a greater proximity of people to wildlife and to other people.

Unfortunately, the potential negative effects on *Homo sapiens* of a deteriorating ecosystem go beyond mortality and morbidity. In the past, human societies have been severely destabilized by shifts in climate,[85] and they may be badly destabilized by climate shifts in the future, too: As currently ongoing climate change exceeds thresholds of extent, intensity, and duration,[86] the resulting droughts and higher air temperatures could prompt large numbers of people to move not just within the nation where they reside but also out of that nation.[87] Estimates of the magnitude of

---

[80]PEW RES. CTR., PUBLIC AND SCIENTISTS' VIEWS ON SCIENCE AND SOCIETY 51,102 (2015).

[81]*E.g.*, Marlieke E. A. de Kraker et al., *Mortality and Hospital Stay Associated with Resistant STAPHYLOCOCCUS AUREUS and ESCHERICHIA COLI Bacteremia: Estimating the Burden of Antibiotic Resistance in Europe*, 8(10) PLOS MED., Oct. 2011, at 1, 5–6; THE REVIEW ON ANTIMICROBIAL RESISTANCE, TACKLING DRUG- RESISTANT INFECTIONS GLOBALLY: FINAL REPORT AND RECOMMENDATIONS 4, 11–12 (2016).

[82]Derek MacFadden et al., *Antibiotic Resistance Increases with Local Temperature*, 8 NATURE CLIMATE CHANGE 510, 511, 513 tbl. 1 (2018); Kate E. Jones et al., *Global Trends in Emerging Infectious Diseases*, 451 NATURE 990, 991 Fig. 1, 992 tbl. 1, [994] (2008). *Accord*, N. Bruinsma et al., *Influence of Population Density on Antibiotic Resistance*, 51 J. ANTIMICROBIAL CHEMOTHERAPY 385, 387–89 & tbls. 2 & 3 (2003) (using data from samples of residents of three cities in different Western nations; comparing rates of antibiotic resistance across the cities; finding, without a multivariate analysis, that the rates were generally highest in the most densely populated city and were generally lowest in the least densely populated city).

[83]MacFadden et al., *supra* note 82, at 511, 513 tbl. 1; Sarah F. McGough et al., *Rates of Increase of Antibiotic Resistance and Ambient Temperature in Europe: A Cross-National Analysis of 28 Countries Between 2000–2016*, BIORXIV, Sept. 12, 2018, at [1], [10], [16] tbl. 1, https://www.biorxiv.org/content/early/2018/09/12/414920.full.pdf.

[84]Note 59 and accompanying text in *supra* Chap. 1.

[85]REID A. BRYSON & THOMAS J. MURRAY, CLIMATES OF HUNGER (1977); Jun Yin et al., *Climate Change and Social Vicissitudes in China over the Past Two Millennia*, 86 QUATERNARY RES. 133, 142 (2016).

[86]Robert McLeman, *Thresholds in Climate Migration*, 39 POPULATION & ENV'T 319, 333 (2018).

[87]Michael Berlemann & Max Friedrich Steinhardt, *Climate Change, Natural Disasters, and Migration — A Survey of the Empirical Evidence*, 63 CESIFO ECON. STUD. 353, 377 (2017) (review of research). *See also* Raphael J. Nawrotzki et al., *Do Rainfall Deficits Predict U.S.-Bound Migration from Rural Mexico? Evidence from the Mexican Census*, 32 POPULATION RES. & POL'Y REV. 129, 135, 139, 150 (2013) (finding that deficits in rainfall in already dry areas of Mexico promoted migration from Mexico to the United States). Weather-caused environmental problems, however, more often generate long-distance migration within a country than migration between countries. David J. Kaczan & Jennifer Orgill-Meyer, *The Impact of Climate Change on Migration:*

such international migration flows vary because the estimates depend on the methodological choices of investigators.[88] A rough idea of this magnitude may nevertheless be obtained from the estimate that all drivers of international migration generated 81 million cross-nation migrants during 2000–2010[89] and from the estimate that the climate-instigated increase in international migrants during 2010–2020 was between 8.6 and 12.8%.[90] Multiplication using these estimates indicates that, relative to 2000–2010, climate change added between seven million people and ten million people to cross-nation migration during 2010–2020. Such a number is sizeable, of course, and it is likely to involve mainly people who relocate from developing to developed nations[91] due to the impact that climate change has on agriculture in developing nations.[92]

What sociological consequences would this migration have had? Migration between nations, according to available research, brings about social conflict in the areas receiving the migrants to the degree that the culture of the immigrants differs from that of native inhabitants,[93] and it strengthens internal social divisions in the receiving areas as well as in the sending areas.[94] The foregoing effects, notably, would probably have been on top of increases in social disorder caused directly by the hotter

---

A Synthesis of Recent Empirical Insights, 158 CLIMATIC CHANGE 281, 284, 288, 294, 296 (2020). *See also* Fernando Riosmena et al., *Climate Migration at the Height and End of the Great Mexican Emigration Era*, 44 POPULATION & DEV. REV. 455, 463–64, 478, 480 (2018) (analyzing data on Mexico from the 2000 and 2010 censuses of that country; limiting the data analysis to local areas having a population below 15,000; pointing out that emigrants from these areas to another nation usually go to the United States; finding that climate changes reduced emigration to another nation from "most" of these areas but increased emigration to another nation from (i) low and high socioeconomic areas and (ii) areas having a history of migration; concluding that climate problems tend to produce domestic migration and migration over short distances); Raphael J. Nawrotzki et al., *Domestic and International Climate Migration from Rural Mexico*, 44 HUM. ECOLOGY 687, 689, 695 tbl. 5, 696 (2016) (analyzing data on members of households in rural areas of Mexico during the period 1986–1999; finding that two out of three measures of high temperatures were associated with emigration from these areas to the U.S.; finding, too, that increases in the number of "very wet" and "extremely wet" days reduced the odds of emigration from these areas to the U.S. but did not affect the odds of migration within Mexico).

[88]Michel Beine & Lionel Jeusette, *A Meta-Analysis of the Literature on Climate Change and Migration* 18–19 (Munich Soc'y for the Promotion of Econ. Research, CESifo Working Paper No. 7417, 2018).

[89]Guy Abel & Nikola Sander, *Quantifying Global International Migration Flows*, 343 SCI. 1520, 1521 Fig. 3 (2014) (reporting that the number of international migrants in 196 nations was 39.9 million during 2000–2005 and 41.5 million during 2005–2010).

[90]Mathilde Maurel & Michele Tuccio, *Climate Instability, Urbanisation and International Migration*, 52 J. DEV. STUD. 735, 749 (2016) (analyzing data on more than 200 nations for 1960 to 2000).

[91]*Id.* at 749.

[92]Chiara Falco et al., *Climate Change and Migration: Is Agriculture the Main Channel?* 2, 11–12, 20–21 (Centre for Research on Energy & Envtl. Econ. & Pol'y, Working Paper No. 100, 2018).

[93]Note 44 and accompanying text in *supra* Chapter 1.

[94]Thomas Faist, *Cross-Border Migration and Social Inequalities*, 42 ANN. REV. SOCIOL. 323, 337–38, 340–41 (2016).

air temperatures that accompany human-caused climate change.[95] The affected societies would have been stressed, in other words, and their ability to operate smoothly would have been hampered.[96]

Put simply, environmentally oriented scholars, especially in the field of law, must address the population problem. They cannot expect economic growth to counter the deleterious environmental impact of the increasing number of human beings, because it will not do so.[97] Instead, they must return to where they were a half-century ago. In the United States, as pointed out in Part 1.1 of Chap. 1, the National Environmental Policy Act—which was adopted in 1970 and is "our basic national charter for protection of the environment"[98]—recognizes population growth as a key determinant of environmental quality.[99]

However, while scholars who are concerned with the environment must once again confront the problem of overpopulation, they will not effectively do so unless they go outside their home disciplines and expand their toolbox with the intellectual resources of other academic fields. For example, scholars who are focused on environmental topics will need to engage with scholars in two humanistic disciplines that can make important contributions to discussions of the population factor, viz., the discipline of ethics[100] and the discipline of theology.[101] They will also need to engage with social scientists such as sociologists who are working to understand how societies interact with nature and handle ecological threats.[102] Because demography is the social science that deals with population structure and processes, environmental scholars will particularly want to become familiar with demographic concepts, data, and quantitative research. Although demographic materials require time to master

[95]Notes 88 to 91 and their accompanying text in *supra* Chap. 1.

[96]*See generally* SAIL VOL. 2, *supra* note 20, at 240–42 (positing three ways in which large-scale migration may alter social life). The sociological impact of migration presumably involves at least some of the factors that determine the incidence of migration. For example, the extent and nature of social networks affects the probability of migration by individuals. Miriam Manchin & Sultan Orazbayev, *Social Networks and the Intention to Migrate* 9–11, 30–31 (Harvard Ctr. for Int'l Dev., Working Paper No. 90, 2018) (studying the impact of social networks on the intentions of residents in more than 150 nations to migrate to another country). If cultures are not the same in the degree to which they promote and support the existence of such networks, the members of some cultures will have higher rates of emigration than the members of other cultures. A high-emigration culture will generate social conflict in a destination area if this culture is dissimilar to the dominant culture of the destination area.

[97]Graeme S. Cumming & Stephan von Cramon-Taubadel, *Linking Economic Growth Pathways and Environmental Sustainability by Understanding Development as Alternate Social-Ecological Regimes*, 115 PROC. NAT'L ACAD. SCI. 9533, 9536–37 (2018).

[98]40 C.F.R. § 1500.1(a) (2020).

[99]Notes 8 to 10, and their accompanying text, in *supra* Chap. 1.

[100]*Supra* note 2 and its accompanying text.

[101]See, *e.g.*, JOHN MCKEOWN, GOD'S BABIES (2014); ABDEL RAHIM OMRAN, FAMILY PLANNING IN THE LEGACY OF ISLAM (1992).

[102]Sophie Peter, *Integrating Key Insights of Sociological Risk Theory into the Ecosystem Services Framework*, 12(16) SUSTAINABILITY art. 6437, at 1, 12–15 (2020).

because they are statistical as well as technical, they can inform and make indispensable contributions to environmental scholarship generally and to environmental-law scholarship specifically.

Unfortunately, law school professors are still not comfortable with proposals to use social science in their work,[103] a situation that extends to law school professors who specialize in environmental law.[104] The situation appears to be at least partly explainable by the marked differences that exist between social science disciplines and the field of law in cognitive approaches and investigative methods.[105] On the other hand, the incorporation of demographic materials into environmental law is now supported by a heightening of interest in the social sciences by law school professors and practicing lawyers.[106] While scholarship originating in the profession of law that employs the social sciences has in the past relied mainly on social science theory,[107] it is increasingly empirical,[108] and with the passage of time, it will be able to take advantage of the constantly expanding body of sophisticated numeric information in the social sciences. Environmentalism and environmental law scholarship, therefore, can continue to ignore demographic concepts, data, and quantitative research only if they are willing to proceed without a social science that illuminates a key player in Earth's ecosystem, and that will provide still-greater insight into this player in the future.

---

[103] Frans L. Leeuw, *American Legal Realism: Research Programme and Policy Impact*, 13(3) UTRECHT L. REV. 28, 29 (2017).

[104] Robert L. Fischman & Lydia Barbarsh-Riley, *Empirical Environmental Scholarship*, 44 ECOLOGY L.Q. 767, 768, 806 (2018).

[105] Howard Erlanger et al., *Is It Time for a New Legal Realism?*, 2005 WIS. L. REV. 335, 336–37 (2005).

[106] Shari Seidman Diamond, *Empirical Legal Scholarship: Observations on Moving Forward*, 113 NW. U. L. REV. 1229, 1229–30, 1231 n.10 (2019); LEEUW & SCHMEETS, *supra* note 49, at 237. For an illustration, see Clare Huntington, *The Empirical Turn in Family Law*, 118 COLUM. L. REV. 227 (2018).

[107] Douglas W. Vick, *Interdisciplinarity and the Discipline of Law*, 31 J. L. & SOC'Y 163, 189–90 (2004).

[108] LEEUW & SCHMEETS, *supra* note 49, at 237. *Cf.* Lee Epstein & Gary King, *The Rules of Inference*, 69 U. CHI. L. REV. 1, 6–10, 15–16 (2002) (concluding that the empirical research published in U.S. law journals is of poor quality).

Index

L. D. Barnett, *Demography and the Anthropocene*,
SpringerBriefs in Environment, Security, Development and Peace 35,
https://doi.org/10.1007/978-3-030-69428-9

About the Author

Larry D. Barnett received a B.A. *summa cum laude* from the University of California at Los Angeles; an M.S. from Oregon State University; a Ph.D. from Florida State University; and a J.D. *with honors* from the University of Florida. He was on the full-time faculty of the Widener University Delaware Law School from 1978 to 2013, and is currently Professor Emeritus. During academic year 1979-80, Professor Barnett was a Fulbright Scholar at the University of Leiden in The Netherlands. In the field of population studies, Professor Barnett founded *Population Research and Policy Review*, a journal in demography that has been published since 1982, and served as editor-in-chief for its first eleven volumes; he also authored the book *Population Policy and the U.S. Constitution* (1982).

Outside population studies, Professor Barnett has written several books on the sociology of law—*Societal Agents in Law: A Macrosociological Approach* (2019); *Societal Agents in Law: Quantitative Research* (2019); *Explaining Law: Macrosociological Theory and Empirical Evidence* (2015); *The Place of Law: The Role and Limits of Law in Society* (2011); and *Legal Construct, Social Concept: A Macrosociological Perspective on Law* (1993, 2010). He is currently finishing a book titled *Societal Stress and Law*.

L. D. Barnett, *Demography and the Anthropocene*,
SpringerBriefs in Environment, Security, Development and Peace 35,
https://doi.org/10.1007/978-3-030-69428-9

Address:
Email: ldbarnett@widener.edu
Website: https://delawarelaw.widener.edu/current-students/faculty-directory/faculty/210/ and https://works.bepress.com/larry_barnett/

About this Book

The environmental movement and environmental law scholarship devote little attention at the moment to the size and growth of the human population. The present work attempts to counter this neglect. To do so, the work includes original graphs that show the magnitude of human population size and growth since 1920 in the world as a whole and in the United States, a country that has an exceptionally large ecological footprint. The work then assembles credible evidence that ties the increasing number of people to ecosystem deterioration and the societal consequences of that deterioration. Given the negative ecological effects of overpopulation, the work considers the hypothesis that the current disregard of population pressures by U.S. environmentalists reflects the thinking of Americans in general. Analyses of data from national sample surveys of American adults reveal that concern with the environment predicts concern with population growth but that the two concerns are not strongly linked. Government in the United States, therefore, is unlikely to take steps primarily intended to lower domestic childbearing, which has been a key source of the expansion of the American population. However, even if steps were taken to reduce fertility and/or immigration, the findings of quantitative social science research on regulatory law and government policy indicate that the steps would not have a substantial, lasting impact on the targeted activity. Social science research nonetheless suggests an indirect way by which law/policy can reduce childbearing. The discussion underlines for environmental scholars the importance of studying their subject in a multidisciplinary, collaborative setting.

[*Keywords:* Anthropocene, attitudes, demography, ecosystem, effectiveness of law and government policy, human fertility, immigration, overpopulation, structural-functionalism theory]